AFRICANIZED HONEY BEES IN THE AMERICAS

May all your bees be
non-AHB's
Best Beekeeping
Dewey M Caron

DEWEY M. CARON
Department of Entomology & Applied Ecology
University of Delaware
Newark, DE 19716

Dedicated to the many beekeepers of the Americas who have shared with me their apiaries and experiences with the Africanized honey bee.

Published by The A.I. Root Company, Medina, OH USA

Mention of a pesticide or proprietary product by the author does not constitute recommendation or endorsement by the author or the A.I. Root Company.

Printed in the United States of America.

Africanized Honey Bees In The Americas
ISBN 0-936028-14-9
2001

TABLE OF CONTENTS

PREFACE vii

CHAPTER 1 **THE WORLD HEARS ABOUT AFRICANIZED BEES** 1
What is the "killer" or Africanized bee? ◆Origin ◆The killer bee ◆The early 1970s ◆Perceptions

CHAPTER 2 **INTRODUCTION AND SPREAD OF THE AFRICANIZED BEE IN THE AMERICAS** 11
How many? ◆Spread of the Africanized bee in Brazil ◆It reaches northern South America! ◆Spread southward in South America ◆Central America & Carribean ◆Africanized bees in the U.S. ◆Early arrival of Africanized bees in California ◆Texas ◆Arizona and New Mexico ◆The main population reaches California ◆Spread eastward ◆Have they slowed down? ◆Eventual distribution in North America

CHAPTER 3 **THE WORLD OF BEES** 27
Subspecies or races of bees ◆*Apis mellifera scutellata* ◆Other African races ◆Why introduction? ◆Previous introductions of bees ◆Why did the African bee succeed?

CHAPTER 4 **WHAT'S IN A NAME?** 39
Morphometric analysis ◆Protein analysis ◆DNA analysis ◆mtDNA analysis ◆DNA analysis ◆So the answer is... ◆Identification in the field

CHAPTER 5 **MYTHS AND BARRIERS** 51
Myths ◆Bee barriers ◆Barrier in Panama ◆Genetic barrier ◆Mexico barrier ◆Revised barrier plan ◆Canadian barrier ◆Barriers revisited

Chapter 6 First Impressions and First Encounters 65

Panama discovers Africanized bees ◆Beekeeping in Panama ◆First experience ◆Beekeeping in the interior ◆Juan Malivern ◆Hermann Gnaegi ◆Penonome school visit ◆Africanized bees in Penonome ◆Changes at McClymonts ◆Other first experiences

Chapter 7 Africanization of Panama 79

Stinging incidences ◆Panama's first sting death ◆Quebrada Bonita ◆National Commission ◆Panama Canal response ◆Africanized bee impact on beekeeping ◆Changes ◆Surviving with Africanized bees

Chapter 8 Beekeeping in Costa Rica 93

Invasion of Costa Rica ◆Arturo Salazar ◆Defensive bees ◆Visit to the Perez brothers ◆Stinging accidents in Costa Rica ◆Beekeepers tour Costa Rica ◆1994 tour ◆Studying the process of Africanization ◆Central valley Africanization ◆High elevation studies ◆Comparison of elevational differences ◆Changes in Costa Rica ◆Paco Vasquez, master beekeeper ◆F_1 stock in Costa Rica ◆Pollination

Chapter 9 Beekeeping in Bolivia 111

Africanized bee arrival ◆A tropical beekeeper—Francisco Kempff ◆Isaac Nuñez ◆National honey festival ◆Tropical beekeeping in the Yungas and Chapare ◆Beekeeping in the Bolivian high Andes ◆Oscar Arispe ◆Mizque ◆Criolla versus Africanized bee ◆National Beekeeping Congress

Chapter 10 Africanized Bees in Other Countries 129

Belize ◆Marijuana defoliation and honey bees ◆Southern Belize ◆Brazil ◆Selection of less defensive more productive bees in Brazil

◆Success with Africanized bees ◆Peru ◆Trinidad & Tobago ◆Venezuela ◆Stinging incidents

CHAPTER 11 MAKING A LIVING WITH AFRICANIZED BEES **147**

Testing for defensiveness ◆Toxic reactions ◆Reducing defensive behaviors ◆Why are Africanized bees so defensive? ◆Dealing with defensive colonies ◆Management for honey production ◆Pollen collection

CHAPTER 12 THE TROPICAL AFRICANIZED BEE **165**

The American revolution ◆The $64,000 question ◆Environmental factors ◆Maintaining productive colonies ◆Swarming ◆Colonizing ◆Defense ◆Is it a subtropical bee? ◆The ultimate myth—feral bees? ◆A new resource ◆Africanized bees—a success story?◆A different bee

REFERENCES **187**

INDEX **223**

Preface

I find the world of honey bees fascinating. Along with most people I recognize honey bees as beneficial. They supply us with honey and beeswax and freely pollinate our flowers as they go about doing their thing. They may sting, but if we leave them alone they generally will not bother us. There is however a population of bees in the Americas that behave, and people perceive, differently. This book is about that bee.

Honey bees are not native to the Americas through few individuals recognize this. Native Americans had plenty of bees to pollinate flowers for the plant diversity they knew and relied upon for survival. Before colonization of North America by Europeans, the Native North Americans lacked a social bee, other than the seasonally social bumble bees. Native Americans of Central and South America did know and did culture social bees, several species of stingless bees. These people had a long history of culture of some of the larger species.

After honey bees were carried to the Americas by early settlers, the introduced social bees prospered as they found the flora of the eastern U.S. and more temperate areas of the southern hemisphere to their liking. As western man expanded his influence and settled elsewhere in the Americas, honey bees quickly followed. The imported European social honey bee did well except in tropical regions of the Americas.

Like others before them, Brazilian beekeepers sought to improve beekeeping by importation of honey bees from elsewhere. European honey bees, despite several attempts at introduction, were not viewed as well-suited for the tropical regions of Brazil. Brazilian beekeepers naturally looked eastward across the Atlantic to Africa. There were some basic differences in beekeeping and in the bees of sub-Saharan Africa, but at least in South Africa honey bees were highly productive (they held the world's record for honey production!) and were managed in a similar way.

In most instances of exotic species introduction, there is little or no impact on native species following importation of non-natives. This had been the apparent case with importation of European bees into North America. In the case of the African bee into Brazil the story is different. The imported bee from Africa has maintained its own genetic identity and found conditions favorable for it to spread, forcing the earlier exotic European bee to disappear.

Part of the "baggage" the African bee brought encoded in its genetic makeup was hightened defensive behavior. The hoped-for advantage in productivity did not occur—at least not in the way it was expected. The imported bee was a distinct contrast from the stingless native bee and the more docile European honey bees previously introduced. Thus this bee received unfavorable press; the Portuguese language term "abelhas assassinas" (assassin bee) became translated as "killer bee."

The decendents of this introduction, the Africanized bee as I prefer to name it in this book continues to receive heightened media attention. There is something about a "bad" insect, which most people recognize as an icon for beneficial, that is newsworthy. This media attention will only intensify as the population of Africanized bees continues northward spread in California. In my estimation it will become especially newsworthy when the population spreads into Florida.

There have been, and there will continue to be, many contradictory statements about Africanized bees. I hope I have not added to the list with this book. In part everything written or said about the Africanized bee population may well be correct—at least in the context of the author/speaker and the bee itself. It can be very productive; it is favored by beekeepers; it is beneficial to humans, yet also feared by humans, overly defensive; and it may be extremely difficult to manage to our benefit. It is a contradiction to itself.

This book on Africanized bees represents many personal experiences with the bee itself, many hours of conversations with and much time spent listening to beekeepers and others who also have experienced the bee first-hand, and organization of the extensive literature—primarily scientific, but also some of the popular press—written about the bee. I was inspired by earlier "killer bee" books by Mark Winston (*Killer Bees* published by Harvard Press) and Greg Flakus (*Living* with *Killer Bees* by Quick Trading Company). Both are excellent presentations and should be read, or reread, with this book for a complete appreciation of Africanized bees.

This book grew out of many presentations about Africanized bees that I have given to beekeeper and other groups. The many, many audience questions during and following such presentations have helped shape what I cover here. I am not sure I have yet encountered anyone who has not heard something about "killer bees." Conversations with beekeepers, friends and new acquaintances often include something about the killer bee. I haven't tired of talking—or writing—about them; and it seems there is no end to questions about them.

The first portion of the book (Chapters 1–5) are of an introductory nature. In Chapter 1, I explain how the population we know as Africanized bees came to be. I also discuss how the world beyond the beekeeper heard about Africanized bees. I include also a section on perceptions about bees, particularly perceptions about the Africanized bee.

Chapter 2 explores the spread of the Africanized bee in the Americas. It concludes with a short section and some speculation on the eventual final distribution of the population. Chapter 3 is a look at the context and extent to which humans have moved populations of honey bees as I attempt to answer two important "WHY" questions in—why was introduction of defensive bees from Africa to Brazil attempted, and why did the introduced bee succeed in colonization of the Americas far beyond anyone's expectations?

Chapter 4 was one of the more difficult chapters to write. What exactly is the bee we label "Africanized bee" and what should it most properly be called? I dislike using the "K" word as most of the general public and much of the press call it. Is the bee a hybrid as the term Africanized would seem to indicate? As I explain there is no definite answer as there still is considerable controversy regarding the mixing of genetic material. The bee is, however, increasingly viewed as largely derived from African origin and not a hybrid of African and European races. I prefer in calling it Africanized, however, as that is a widely-recognized term for the bee. Doing so, however, does not imply that it is a hybrid. Readers may disagree with my assessment—and the general public will likely persist in calling it killer bee.

Chapter 5 was the most enjoyable chapter to write and it is one reviewers liked to read. It grew out of talks I gave on Africanized bees and in listening to colleagues—other bee scientists and beekeepers—who had first-hand experiences with the Africanized bee population. It is the myths and [mis]perceptions about Africanized bees that I hope to clarify with this book so there is always some danger in writing them down else they become accepted as fact. I do invite you, the reader, to think about and even list other myths and perceptions you have or have encountered after reading this chapter.

Chapters 6–11 are a tour of several countries and a look at beekeepers coping with Africanized bees. My first encounter with Africanized bees was in Panama in 1982. I had the opportunity to be on hand as the second Africanized bee swarm was captured in Panama and I returned to see the changes the Africanized bee brought to Panama in some 15 return visits over the past 17 years. Chapters 6 and 7 are both about Panama.

Chapter 8 is the situation in Costa Rica. I have been able to visit Costa Rica and bring beekeepers and students to Costa Rica for extended periods and trips. Dr. Marla Spivak did a series of important studies on Africanization of Costa Rica. I am grateful for her information that I have attempted to share with you—any errors in interpretation are mine and are not due to her careful work. The beekeeping situation with Africanized honey bees is different in Costa Rica. Understanding Africanization in Costa Rica and their solution is important for U.S. beekeepers as it may point the direction for our own country.

Chapter 9 is another piece of the puzzle to our understanding of what is the Africanized bee—it describes the situation in Bolivia. Through my in-laws and very patient wife, Nieves, I have had a chance to spend a fair amount of time with beekeepers and studying the Africanized bee in Bolivia. The same bee is different depending upon environmental conditions related to elevation in Bolivia. Africanized bees at higher elevations are much more manageable and ameanable to beekeeping. It is an encouraging bit of information to temperate-area U.S.beekeepers.

Chapter 10 looks at the bees in 5 other countries. Each has a different experience and each has attempted its own solution to Africanized bee colonization. The bee varies so also does the response—as with beekeeping with European bees there is no one right way or a wrong way to keep Africanized bees. But some ways are better than others and this look at what is happening elsewhere in the Americas is offered for the lessons we might take from the experiences of others.

The last two chapters are a look at defensiveness and productivity to further help beekeepers understand and profit from Africanized bees. The Africanized bee is not all good nor is it all bad—both chapters give a roadmap of how beekeepers can manage Africanized bee recovery programs.

For a complete understanding of what we know, readers should begin with three major reports on Africanized bees—the 1988 Ohio State Conference on Africanized Bees and Bee Mites (proceedings edited by Needham, Delfinado-Baker, Page and Bowman), *The "African" Honey Bee* (published by Westview Press 1991, edited by Spivak, Fletcher and Breed.) and a soon-to-be-published (A. I. Root Co.) *Proceedings of the Second Africanized Bee and Bee Mite Conference* held in Tucson, AZ, in April 2000. Extensive (but by no means complete) references for each of the chapters are listed in a separate section following Chapter 12.

There are many people to recognize for their contributions to the content and ideas of this book. First and foremost I would like to thank the beekeepers of the Americas who shared their Africanized bee apiaries and stories with me. Several colleagues saw portions of the drafts and offered excellent suggestions and constructive criticism. I thank them and all the researchers referenced for their input into the Africanized bee story. I am responsible for interpretation of their experiences and their information and I trust I have done so fairly and correctly.

I have borrowed illustrations and visuals from a number of colleagues. I thank them for the permission to do so and appreciate their willingness to contribute to the book. I also wish to acknowledge the many beekeepers who allowed me to take photos of their bees and beekeeping—unfortunately I am able to include only a very few of those in this book but hope the ones I have included are instructive.

I wish to especially acknowledge Ann Harman who patiently worked with me to get the manuscript in final form for the publisher and who offered innumerable constructive suggestions for improvement of this edition. I finally thank my patient children, Michael, Mark and Deborah, who saw me off countless times to countless places (and were able to join me for a few of those trips). And a special thank you for my wife, Nieves—for allowing me the time to put this all together.

I encourage you, the reader, to experience and learn about Africanized bees first-hand. Experience may be the best teacher of all.

Africanized Honey Bees

in

The Americas

1

The World Hears About Africanized Bees

> "A strain of honeybee not yet present in North America seems likely to enter that continent from the south, if its spread is neither hindered nor helped through human agencies. This strain, now rapidly extending its range in South America, has both objectionable and dangerous attributes. Because of its unprovoked mass stinging and because of frequent swarming and absconding, the Brazilian bee is dangerous to people and animals and is difficult to manage. It has already brought about drastic changes in the beekeeping industry in the areas it occupies in South America." (From: 1972 NAS Committee report)

Our earliest primate ancestors hunted feral (wild) honey bee nests. Then, as today, the sweet and medically-important stored honey reserves and the beeswax in which it is stored were valuable commodities. Humans have only been intensively managing colonies of honey bees (beekeeping) in man-made domiciles (hives) for a few hundred years. Swarms that leave managed hives still survive and prosper in the wild, testimony to the small domain we have over bees.

How people perceive honey bees depends upon past experiences including what they have read or heard about bees. The general public today has little contact or direct knowledge of honey bees. However, many individuals have heard about and even know the basics about the Africanized or "killer"[1] bee.

The procedure to manage colonies of honey bees, or beekeeping, has changed most dramatically over the last 150 years. Colony care begins with understanding and adapting basic honey bee biology to a convenient handling system. Of all the innovations, discovery of the principle of "bee space" led to development of a hive with removable combs for easy study and manipulation utilizing the natural biology of the bee nest. Subsequent development of other beekeeping tools (such as the bee smoker, honey

---

[1]Many names have been applied to the population of honey bees carried to Brazil from Africa. I prefer to use "Africanized bee" as explained on the next page. See Chapter 3 for a more thorough discussion of names.

extractor, queen mailing cage) and improved management systems, have resulted in an increase in quantity and quality of honey and other harvestable products of the honey bee.

Thus, humans have come to rely upon bee products to help feed and medicate a growing human population. The list of commodities obtained from a bee hive is considerable—sweet, delicious honey, bee-collected pollen, beeswax, bee venom for therapy of arthritis-related human conditions, royal jelly, and propolis are all natural, pure, readily-harvestable bee products. The greatest benefit of all from honey bees however is use of managed bee colonies in planned pollination of over 100 food, seed, and fiber plants.

How is it that the beneficial honey bee, the indispensable pollinator, our insect helper and provider of a wide variety of products has come to be labeled a "killer bee," a creature dangerous to people and animals?

What Is the "Killer" or Africanized Bee?

Basically the Africanized bee is a population of honey bees with a bad reputation! Africanized or killer bees behave differently from European honey bees (the more widely-distributed population or common honey bee). Defensive behavior (i.e. stinging) is a well-known difference but Africanized bees differ also in nesting biology, their tendency to swarm, absconding behavior, pollination effectiveness, foraging behavior, honey-storing characteristics, as well as many other behaviors. These differences affect beekeeping most directly but they also impact agriculture and the general public.

The rapid and widespread colonization of the Americas by the population of Africanized bees was a totally unexpected consequence of the introduction of an African honey bee subspecies (race)[2] into Brazil. Other honey bee populations had been previously introduced into the Americas with very positive outcomes. What happened with this introduction?

Mark Winston, a Simon Fraser University Professor and leading Africanized bee researcher, in his book, *Killer Bees*, labels the African bee "one of the most astounding success stories in the history of introduced insect species." Tom Rinderer of the USDA Bee Lab in Baton Rouge, LA, another leading researcher on Africanized bees, states that Africanized bees are "...perhaps the most remarkable biological event of the century." Will the Africanized bee "potentially...be the single most severe insect pest in the United States" as Al Dietz, retired apiculture professor from the University of Georgia, predicted in the most recent edition of *The Hive and the Honey Bee?* Are we to realize his prediction that "perhaps 200 or

---

[2]The concept of race and subspecies in honey bees is discussed in Chapter 3.

more persons...[will]...die annually in the United States from massive stinging and 2,000 to 3,000...[will] be hospitalized for the same reason."

Figure 1–1. Colony entrance—how can we tell they are Africanized bees?

Many people have the mistaken impression that the Africanized bee has mellowed because it has hybridized with the more gentle, managed bee colonies introduced earlier from Europe into the Americas. This is, most emphatically, not the case! It is widely assumed that U.S. beekeepers, the most advanced and knowledgeable in the world, will be able to handle the Africanized bee. This, too, may not be true, at least not for the majority of U.S. beekeepers who live or keep bees where feral populations of Africanized bees will likely establish permanent nests.

The behavior of a colony of honey bees depends upon inherited characteristics (genetic makeup) as influenced by environmental factors. Beekeepers willingly interact with their managed bee colonies and have certain expectations of them. So, too, the general public has expectations of what honey bees should be, even if they have minimal contact with them. Beekeeper and general public expectations are not necessarily at odds, even though their experiences and interactions may be different.

We do not know yet how extensively the Africanized bee population will spread in North America nor the eventual impact it might have! We do know about largely negative impacts it has had on South and Central American peoples and beekeeping. Hopefully the area of colonization of the Africanized bee population will affect only a minority, not the majority, of U.S. beekeepers and general public. Predictions, along with perceptions and expectations, vary widely.

Origin

The Africanized bee is a population of more or less distinct honey bees. It is the offspring of a small number of queens of a large population (race) known as the African bee or *Apis mellifera scutellata*[3] that occurs naturally in parts of Africa. These queens, adapted to living in tropical Eastern Africa for centuries, were carried to Brazil as a first step in an attempt to improve beekeeping in tropical regions of the Americas.

In its natural range, the African honey bee is a honey bee ecotype naturally adapted for the tropical climate and environment of parts of Africa. As with other populations of honey bees, it displays a considerable amount of genetic variation. Among a variety of traits, the African bee race was known to be "bad tempered," a defensive bee, but it was also considered to be an excellent honey producer.

The African bee queens brought to Brazil were to be used to initiate a honey bee breeding program to develop an ecotype better suited for the American tropics. The objective was to incorporate the productivity of the African bee with the more gentle nature of the European-race bees introduced earlier to South America. The African bees escaped before initial efforts to evaluate their potential were completed. Following accidental release, their genetic material was openly distributed with little effort to breed for gentleness.

Then the unexpected happened! The African bee genetic material was apparently not diluted by mixing with the larger population of European bees and African bee genetic material replaced European (see Chapter 4 for a more extensive discussion). Natural selection favored the genetic makeup of the newcomer. African bees apparently had more adaptive behaviors and their biology provided for a selective advantage in the tropical environment. Subsequently, this bee "Africanized" thousands of managed European bee colonies producing an unbelievably large feral population that has colonized all of the Americas except Chile, Canada and a few islands in the Caribbean.

The Killer Bee

The defensive behavior of the Africanized bee, dubbed "killer bee" by the popular press, has captured the attention of the general public. Because the Africanized bee has retained its original defensive behaviors, stinging incidents have led to injury or death of animals and humans who happen to be in the vicinity of one of the disturbed feral nests or managed colonies of a beekeeper. Such incidents occur when the bees sting in response to a disturbance. Victims usually are simply in the wrong place at the wrong time.

---

[3]Initially it was incorrectly labeled *Apis mellifera adansonii* (see Chapter 3).

The defensive response of Africanized bee colonies is often exaggerated and prolonged. Such pronounced reactions might occur in any colony of honey bees but are only rarely experienced with European bees. Until beekeepers learn to manage colonies differently, and the general public treats this defensive population differently, stinging incidents with "killer bees" are likely to continue.

In 1990 the Africanized honey bee arrived in the U.S. Various efforts to halt its spread in the Americas and prevent the Africanization of managed European colonies have so far proven inadequate, though our understanding of honey bees and the biology of this tropical bee ecotype has greatly advanced since we first heard about them. Will such knowledge enable us to finally halt the spread of the bee? Or, perhaps, has the Africanized honey bee finally reached climatic conditions where it will not prove superior to the European bee?

The Early 1970s

Two events in the early 1970s changed the obscure world of honey bees and beekeeping. One was the awarding of the Nobel Prize to Professor Karl von Frisch for his discovery and presentation of honey bee dance language to the world. The recognition of von Frisch, and two other colleagues, helped legitimize the growing biological specialty of ethology—the study of animal behavior.

With his careful, highly innovative research, much of it performed during the turmoil years of Nazi Germany and World War II, von Frisch opened the eyes of the world, and fellow scientists, to the behavior by which honey bees inform sisters of the precise location, identity and richness of discovered food sources. With his discoveries we could now appreciate and interpret the elaborate and elegant repeated patterns of bee movement called "dancing."

A second early '70s event, equally as momentous, was the release of a National Academy of Sciences (NAS) study on the accidental introduction of a defensive race of honey bee into the Americas.[4] This was the introduction of the bee known as the "killer" or "bad bee of Brazil," as TIME magazine headlined in the September 24, 1965, edition.

The U.S. National Academy of Sciences team was a blue-ribbon team of 12 top bee scientists and ecologists. They spent three weeks in Brazil during November 1971 interviewing beekeepers and bee scientists. They examined the growing literature of scientific studies on the Brazilian bee (virtually all in Portuguese,of course) and spent time looking at bees, even developing a method to scientifically measure defensiveness.

---

[4]Source of quote at beginning of chapter. Hereafter this 1972 report is simply referred to as NAS committee report. For information on sources quoted and consulted see References following Chapter 12. This report is referenced as National Academy of Sciences 1972.

PB-211 952

Committee on the African Honey Bee

National Research Council

prepared for
Agricultural Research Service

JUNE 1972

Distributed By:

NTIS

National Technical Information Service
U. S. DEPARTMENT OF COMMERCE
5285 Port Royal Road, Springfield Va. 22151

Figure 1–2. Cover page of 1972 NAS report on Africanized bees.

Some committee members traveled to the north of Brazil and others journeyed southward to cover as much of Brazil as possible. Both groups met with beekeepers and both asked similar questions. The committee members who went to the north heard about, and repeatedly observed, highly defensive bees. Here beekeeping was declining with bee colonies more likely to swarm and less productive. Committee members traveling to the south of Brazil found more beekeepers prospering and colonies were less defensive. They reported increasing rather than declining interest in beekeeping.

The Africanized bee (called Brazilian bee in this report) was not well-known prior to the NAS committee report. Although some scientists had heard of it, few had experienced Africanized bees first-hand. Dr. Warwick Kerr, the bee geneticist that personally hand-carried the African bee genetic material into Brazil, had written little about the spreading population. While visiting the U.S. to attend the 21st Apimondia Congress at the University of Maryland in 1967, he briefly discussed defensiveness and his early mating experiments with the introduced bees but gave little indication of the growing problems being experienced by beekeepers.

S. E. (Mac) McGregor, then at the Tucson bee lab but later to become head of the USDA bee research program, was in the audience and listened to Dr. Kerr. Two years later he visited beekeepers and researchers while attending the First Brazilian Apicultural Congress in Santa Catarina state (in the south of Brazil). He returned "thoroughly convinced that the United States should give serious consideration to this problem."

The general public as well as beekeepers had heard little about the accidental introduction. TIME magazine had two articles on the bee. The first (September 24, 1965), based on a report from a Brazilian newspaper about a stinging incident, used the term "killer bee" simply as a direct translation of the Portuguese name for the bee, "abelhas assassinas." (In Spanish speaking countries the bee is more frequently called "la abeja mala," [the bad bee], or "las abejas extranjeras" [the foreign bees]).

National Geographic had their first article, entitled "Those Fiery Brazilian Bees," in the April 1976 issue. The text began "The bees came on us like a squall...." The lead photo showed the photographer with bees flying about her veil and camera. After the lead and initial photo, the remaining photos and text were relatively tame. We really knew the bees arrived, however, when the *Reader's Digest* article "The Truth About Those Killer Bees" was published the same year.

Why were stories of honey bee stinging incidences in Brazil covered by such well-known publications? Brazilian bee scientists Lionel Gonçalves, Daniel Stort and David De Jong label the years 1964–70 as a period when "an intensely negative campaign was being waged against Africanized bees both in Brazil and abroad." The authors state that the local news media, for "politically motivated reasons...sensationalized...[and] ...greatly exaggerated" the bee situation, resulting in a "negative reaction from the public towards beekeepers" throughout Brazil.

Perceptions

Since these early accounts, there has been a steadily growing amount of information about the Africanized bee. Dave De Jong, in the August 1994 *Bee Culture* magazine, wrote about Africanized bees:

> "The expectations that people in Brazil have concerning the bees have changed. This is true both for beekeepers and the public. They understand that the bees are sensitive insects and deserve respect. The way in which the bees are perceived has become modified as has the reactions of the public to any contact with them. These changes were in great part due to the beekeepers' profound abilities to handle these bees without provoking them.
>
> "In 1970, at the First National Beekeeping Congress, beekeepers were asked to list their principle problems. Excessive stinging placed first. In 1980, at the Fifth Brazilian Beekeeping Congress, this same question was asked again. Stinging problems figured a lame fourth on the list. Today it is not even mentioned.
>
> "What has changed, the people or the bees, or both?"

The public perception of the Africanized bee is not necessarily the same as how beekeepers perceive the bee, or how they necessarily want the general public to perceive the bee. One thing is certain—people have heard about "killer bees." Practically every beekeeper has had the question "What about the killer bee?" posed in one form or another, time and again by family, friends, or acquaintances. It is often the first question asked of a beekeeper.

The number of people who have had an incident or experience with the Africanized bee in countries such as Panama or Costa Rica, where a large feral Africanized bee population now exists, is quite extensive. Native campesinos (subsistence farmers) and people who work in rural areas talk of instances when a swarm passed overhead while they were a long distance from anyone or anything that could have helped them if the bees started to sting. Beekeepers who work around bees all the time know this seldom happens with European bees. It is a common occurrence with Africanized bees.

Stories about honey bees that people in the U.S. tell, hear or read are stories that are often positive with happy endings. The swarm that did not get away, the gentleness of the bees, the beekeeper who came to rescue the citizen in distress, etc. People in countries inhabited by Africanized bees are often relating a negative interaction. Their stories are of the bees that stung and killed an animal, the number of people sent to the medical clinic, of the fear and panic they or the victims felt. These stories are different because the bees are different.

Let me relate one story as an example. In 1988 my son Michael and I had the occasion to visit Copan, an ancient Mayan ruin on the border of Honduras and Guatemala. Copan is a marvelous example of a Mayan city, easily accessible from San Pedro Sula on the Caribbean side of Honduras. It has beautiful temples, a restored plaza and ball field and lots of carved stone columns. The ruins are well-illustrated in the October 1989 *National Geographic*.

On the day of our visit, everyone in Copan (approximately 2000 people live in the present town, about one mile from the ancient one) were talking about the school teacher who was stung to death by bees the day before. She was preparing the noonday meal when her two youngsters, 6 and 4, began screaming in the backyard. She rushed out of her house to find thousands of bees stinging the two boys and the family dogs. She swatted at the bees and pushed the children into the house. She must have then tried to rescue the dogs but sustained a large number of stings. With the exertion, stings, heat and excitement, she collapsed. Her husband and neighbors, hearing the commotion, braved the stinging bees to finally cover her with a blanket.

The teacher was carried to the health center, wrapped in the blanket, but she died shortly thereafter. The two youngsters survived. The two dogs died as did some chickens; the local veterinarian saved a neighbor's pig. That night, neighbors burned the tree, and part of the house, to eliminate the bees. Nobody knows what upset the bees. A tragic story, not the type of story heard frequently with common European bees.

As an extension specialist, I work to change people. I contend, although not all might agree, that humans have not changed the honey bee. When beekeepers began to seriously note disease problems, the defensiveness and low productivity of the German bees initially carried by the colonists to North America, they were ready to try Italian bees being recommended by a few German and Swiss beekeepers. That importation proved immensely profitable. Today North American (and world) beekeepers principally husband Italian bees. The German bee has been forgotten and only a few pockets of wild (feral) colonies of predominately German origin are left in the U.S.

When Brazil attempted to import African bees to crossbreed with the Italian and German bees maintained by their beekeepers, the outcome was unpredictable and certainly not as favorable. This introduction was not the first time, nor is it likely to be the last time, events do not proceed as planned.

There is a biological problem with Africanized bees. Perhaps there is a larger one with beekeeper and general public perception of the biological problem. To answer the question posed by Dave De Jong at the end of his 1994 article, I believe both the bee and people have changed in Brazil. It is people, beekeepers especially, but also the general public, that have changed far more than the bee itself.

✧✧✧✧✧✧✧✧✧

As a lifelong teacher and advocate of the role of extension in transferring appropriate technology, I relate my experiences and the knowledge of what is known about Africanized bees in an effort to expand your knowledge base, change some perceptions perhaps, and allow you to benefit from what I have learned about Africanized bees in the Americas. Although I do not expect you to agree with all that I write, I do hope you might consider these words and if they change your perceptions even a little, I will know it is all worthwhile.

The literature on Africanized bees is very large and not always in agreement. I encourage you to read more of this literature to help you develop an informed, and more realistic view of Africanized bees.

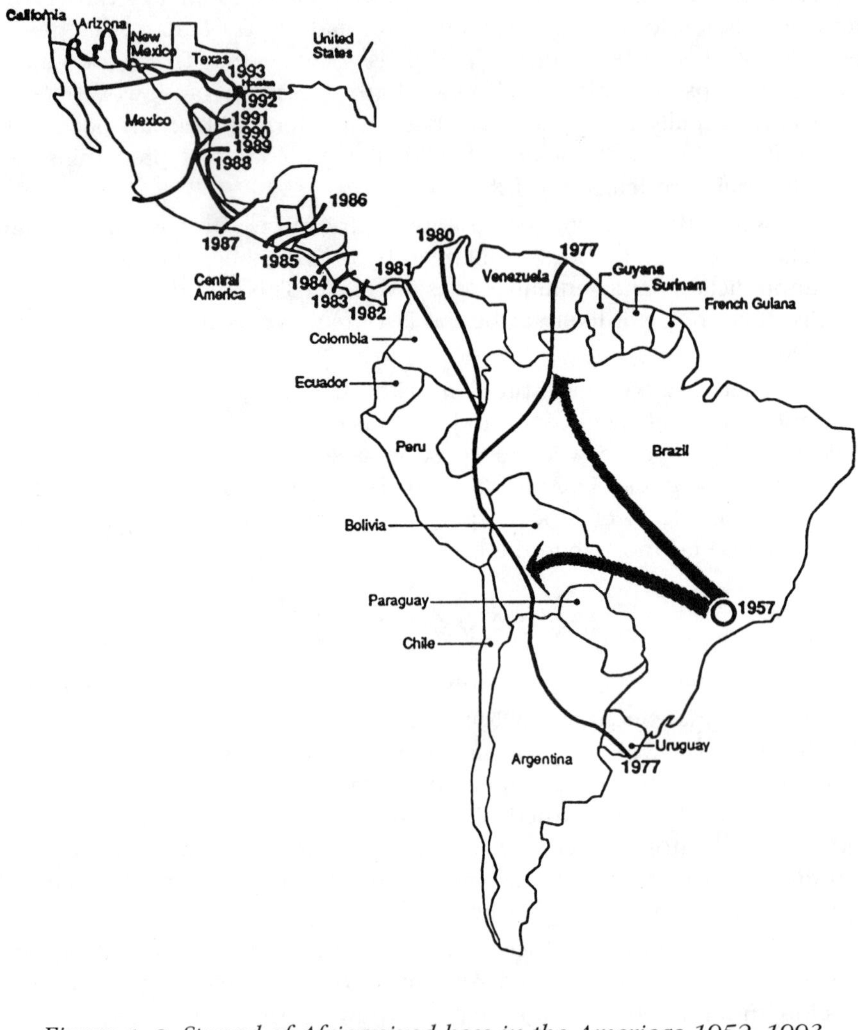

Figure 1–3. Spread of Africanized bees in the Americas 1952–1993. (Source USDA)

2

Introduction and Spread of the Africanized Bee in the Americas

Populations of the honey bee *Apis mellifera* have been carried repeatedly by humans around the globe. Honey bees transported from Europe have done well in North America, Australia and New Zealand and in subtropical regions in the Americas (Mexico and Argentina). European bees have not done as well in Asian, African or American tropical areas. The African bee has found the American tropics to be quite satisfactory.

Beekeepers in Brazil were not satisfied with the German, Italian or Spanish honey bees that had repeatedly been introduced into South America, so they looked to Africa as perhaps a more appropriate source of stock. The Ministry of Agriculture was lobbied to import African bee stock into Brazil. When Dr. Warwick Kerr, a world-renowned geneticist and specialist in native stingless bees, won a Brazilian scientific award in 1956, the cash portion enabled him to go to Africa to evaluate African bee races directly and to select appropriate queen stock for Brazil.

Dr. Kerr's first attempt to ship queens, collected from throughout Africa, was unsuccessful when the queens were sprayed with DDT by a customs official. Thus on his return in 1956, Dr. Kerr personally carried 63 live queens from South Africa to Brazil.

One queen came from Tabora (1200 m [3937 ft] 5°S) in Tanzania (then called Tanganyika). The remainder of the queens were obtained from two well-known South African beekeepers: W. Crisp, a queen breeder who supplied most of the queens, and E. Schnetler, the holder of the world record in honey production. Both South African apiaries were a few miles south of Pretoria, South Africa (1500 m [4922 ft] 26°S).

Forty-seven queens survived introduction into University of São Paulo bee colonies at Piracicaba in São Paulo state. Perhaps these 47 were the survivors of introduction and initial selection since some "which produced the most aggressive progeny were eliminated" according to Polish bee scientist, Hermann Woyke, who spent a sabbatical in Brazil.

The colonies were subsequently moved to a eucalyptus[1] forest belonging to a railroad company in Camaquan, Rio Claro, (São Paulo state) in November 1956. One story says the move was to take advantage of the

[1]Eucalyptus is a good pollen source and November is Spring in Brazil when bees need to collect much pollen.

eucalyptus nectar source but Noel Kempff Mercado, a Bolivian studying at Rio Claro in 1957, would write in 1970 that the bees had become unpopular on campus due to their defensiveness, particularly following their manipulation. Perhaps defensiveness may also have been responsible for their being moved?

It is there where the "accident" occurred! A "railroad employee" removed the double queen excluders from the entrances of 26 of these colonies in early 1957, apparently because he saw pollen loads being removed from workers returning to the colonies. In some versions of the story, queen excluder removal was done by "a well-meaning beekeeper"; in others, "a visiting beekeeper" removed the excluders By whomever and for whatever reason, the consequence of the action was momentous—the colonies swarmed soon thereafter.

How Many?

It is widely stated that the current population of African bees in the Americas originated solely from swarms emanating from these 26 colonies. Has the population, in fact, come from this single accidental release? The story is not consistent and it is widely rumored that it has not been told correctly. (See for example pages 34–39 in *Living* with *Killer Bees* by Greg Flakus.)

The scientific literature includes reports of experiments, apparently with the imported stock, that involved queen rearing and artificial insemination. There were controlled studies comparing African, Italian and German bee colonies in the late '60s. Professor Kerr stated the year following introduction, in his first article about the introduction, that the 26 colonies were "the most prolific, productive and industrious bees that we have seen up to now."

Professor Kerr, in his 1967 Apimondia trip to the U.S., said African queen stock was "distributed widely among beekeepers in southern Brazil." Dr. Orley Taylor, University of Kansas ecologist and a well-respected Africanized bee expert, reports Kerr as personally saying the single queen from Tanzania was "used most extensively for queen rearing" and Nogueira-Neto wrote in a 1964 *Bee World* article that the Tabora queen "left many descendants." It seems likely that, following the accident, there were breeding experiments and distribution of queens reared directly from the African bee stock.

The conventional wisdom, in accord with many studies of the impact of an introduced species on native species, is that most exotic species introductions result in no change in local species abundance or in community structure. This time it was different. Accidental release of African bees, even with the possibility of subsequent further distribution, ranks as an "unexpected consequence" of the release of a small amount of genetic material into a much larger existing population of Italian and German bees managed by experienced beekeepers.

Spread of the Africanized Bee in Brazil

The magnitude and rapidity of the spread of the feral Africanized honey bee population in the Americas has been truly phenomenal. (Figure 1–3) Forty-five years after introduction its range extends from Argentina (approximately 34°S) to five states in the southwestern U.S. from approximately 26°N at Brownsville, TX, to 32–34°N in New Mexico, Nevada, Arizona and California. (Figure 2-4) It lives from sea level to over 3000 m (10,000 ft) in the Andes mountains in South America. The variety of habitats where it has spread and become abundant includes tropical dry forest, savanna, desert, rain forest and cloud forest. This bee has essentially totally replaced European-race honey bees throughout this range.

The initial movement of the bee was relatively slow. For the period 1957–1963, when the Africanized bee reached Rio de Janeiro from São Paulo, spread was calculated as approximately 80 km/yr (50 miles/yr). During this initial expansion it eventually became evident that a feral population of the bee, with the objectionable African characteristic of defensiveness, was becoming established and replacing European bees.

The next five years witnessed a very rapid advance of the Africanized bee population toward the north of Brazil (300–500 km/yr – 186–310 mi/yr). The beekeepers of Brazil's neighbors in Paraguay (1964), northern Argentina (1965), and the tropical region of Bolivia (1967) began to report populations of the bee. (Figure 1–3)

The Africanized bee continued rapid movement northward 250–400 km/yr (150–250 mi/yr) along the coast of Brazil, crossing the mouth of the Amazon River in 1971. Advance was less rapid and less noticeable in the dense Amazonian tropics. By 1972, when the NAS team of scientists from North America visited Brazil (Chapter 1), the bee had spread over an area equivalent to the continental U.S.

It Reaches Northern South America!

The bee arrived in the northern South American countries of Guyana and Surinam during the late 1970s. It spread rapidly along the dry coastal region, as it had in Brazil, but slower inland in wetter, more humid regions. As it spread into the wet forested region of Venezuela, for example, movement was only 160 km/yr (100 mi/yr).

By 1980 the bee was reported throughout Venezuela. It spread throughout the Amazon River basin of Brazil, Bolivia, and Peru. In the eastern tropical area of Peru, the bee inhabited areas up to about 1500 m (5000 ft) elevation in the Andes mountains. Since then it has arrived in high-elevation intermontane valleys of the Andes, though feral populations have not become permanently established at higher elevations in Peru, according to sampling by geographer Robert Kent. In my own recent samples it is clear Africanized bees have successfully invaded such valleys in Bolivia.

Spread of the Africanized bee population west of the Andes Mountains occurred in Peru though the bees did not cross the Andes Mountains unaided; Dr Kent attributed movement to two beekeepers transporting 300 colonies from eastern Peru to the coast. The coastal population subsequently spread northward into Ecuador by 1982. It will likely expand farther southward along the coast and may eventually bring Africanized bees into Chile, though this has not yet occurred.

Spread Southward in South America

Southward expansion of the Africanized bee population has not been as extensive as the northward spread. There has been very little movement southward in more temperate Argentina and Uruguay since 1968. A complex zone of hybridization has developed in this region and the population is reported stable. (Figure 2–1)

Bees in the northern portion of Argentina are Africanized in morphological features, behavior and other measures including nuclear DNA and mtDNA analysis (Chapter 4). The direction of gene flow is from Africanized feral to managed European bees. In the southern portions (below 34°S latitude), bees are European, though pockets of temporary colonization or hybridization may exist, perhaps due to favorable seasonal weather conditions.

In the intermediate transition zone, approximately 32° to 34°S, there is evidence of mixing of African and European genetic material through mtDNA and morphometric measurement analysis. In Uruguay, analysis of body proteins (allozymes) reveals a transition cline north to south. At 30°S (the border with Brazil) there is virtually no European bee protein, while at 35°S (on the Argentina border), more European protein material is present.

The beekeepers of Uruguay believe they have a unique honey bee, a "race…relatively free from genetic influence of Africanized honey bees" that has developed in the Uruguayan environment. Mike Burgett and associates, in a VOCA-sponsored study, collected 53 bee samples for morphometric and mtDNA analysis. Nearly 83% displayed evidence of African genes—53% were classified as hybrid, i.e. they had African mtDNA but European morphometric characteristics. In an analysis 15 years earlier Professor Kerr and colleagues found 2% of the bee population in Uruguay Africanized; Burgett's sample analysis found 30% Africanized.

Frank Eischen, of the USDA bee lab in Weslaco Texas, points out that Argentina and Uruguay are a mirror image of the U.S. The most southern points of Texas and Florida (26°N latitude) are only 80 miles different than the most northern point of Argentina. If the bee establishes a transitional line in the U.S. between 32 to 34°N latitude at the same locations as it has in Argentina and Uruguay (Uruguay lies between 30 and 36°S latitude), it will potentially expand to somewhere between Houston and Dallas in the U.S.

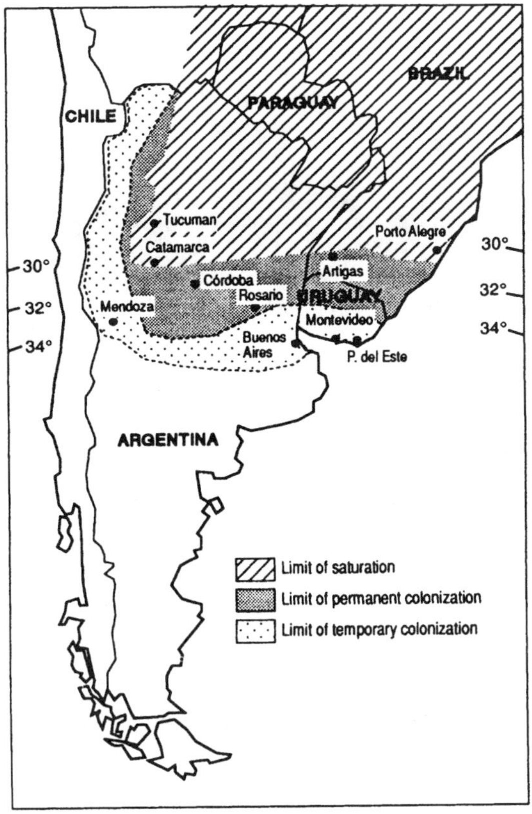

Figure 2–1. Africanized honey bee population limit in Argentina. (From Winston 1992. Original: American Bee Journal 1982)

Central America & Carribean

The Africanized bee invaded the isthmus connecting South America to Central America in late 1981. It was confirmed in Panama in 1982. It spread rapidly northward at a rate of 250–400 km/yr (150–250 mi/yr) along the Pacific Coast of the six Central American republics. (Figure 1–3) Here the climate and annual precipitation of less than 2000 mm/yr (80 inches) was to its liking. The Pacific Coast is the principle region of beekeeping in all six of the Central American countries, though none of the countries previously had large numbers of beekeepers or managed colonies.

The first Africanized bees arrived in Cuidad Hidalgo, in the most southern province of Chiapas, Mexico, in December 1986. They spread rapidly in Mexico. Monitoring a trap line of pulp-pot bait hives (Figure 2–2) located 150 km (93 mi) from the Texas border, Bill Wilson and colleagues from the Weslaco USDA lab found 100% of feral colonies Africanized within two years of initial arrival of the population. The process of colonization was slower in managed colonies. Only 20% of managed colonies in several apiaries sampled in northeastern Mexico were Africanized by 1992; almost all beekeeper colonies showed some Africanization by 1995.

Figure 2–2. Pulp-pot traps used to monitor spread of Africanized bee population. (USDA photo)

The Africanized bee colonized the island of Trinidad in 1979 by flying approximately 8 miles from the South American mainland. It has been able to reach Tobago (1988), Puerto Rico (1994) and the U.S. Virgin Islands (1995) with human assistance. The remainder of the Caribbean islands are so far apparently free of Africanized bee populations.

Africanized Bees in the U.S.

Arrival of Africanized bees into the U.S. was confirmed on October 15, 1990, when a swarm was captured in Hidalgo, Texas, ironically a city with the same name as the community where the bees were first reported in Mexico four years earlier. By the following June, 59 additional swarms or feral colonies were found in the same area of extreme southeastern Texas, mainly in Brownsville and along the Rio Grande River.

The discovery of Africanized bees in Texas, signaling invasion of the main bee population, was not the first time African bee genetic material reached the U.S. Numerous interceptions of swarms and colonies aboard ships have been made, mostly at Gulf Coast or California ports. In one case, a Florida beekeeper retrieved a swarm from the Panama City,

Florida, port and rather than destroy it, hived it in his nearby apiary. By 1999, seventeen Africanized bee intercepts had occurred in Florida, including some in Jacksonville that apparently arrived from Puerto Rico. Florida currently maintains over 500 bait hives to attempt to detect Africanized bees and all intercepted Africanized bees are destroyed.

Another introduction of Africanized bee genetic material into the U.S. was in the form of Africanized bee drone semen shipped to Steve Taber at the USDA bee lab in Baton Rouge, LA. "Many shipments" of semen were used to foster 10–20 colonies that were up to 90% African. Only the F_4 colonies (the most pure African) were destroyed at the end of the experiments. Taber did note that the colonies "were not more aggressive or temperamental than many other bees I had worked with." Harry Laidlaw also received Africanized bee semen in Davis, California, but his material was backcrossed or "genetically washed" so one particular gene, Ac, and "only a very few associated genes" were involved.

In the 1970s while scientists were raising queens in Baton Rouge, LA, Richard Adee, a widely respected U.S. beekeeper, with over 40,000 bee colonies in a dozen states noticed some "problems." His colonies "...were swarming a lot and showing increased aggressiveness." Another beekeeper, one of the largest queen breeders in Louisiana, noticed the same thing. The two "...concluded that African strains that had mixed into the feral bee population around Baton Rouge must have infiltrated their colonies." Adee said requeening took care of his problem; the queen breeder sold his bees to a beekeeper in Missouri.

Steve Taber, who conducted the USDA experiments, dismisses the story. Tom Rinderer, current lab chief at the USDA Lab in Baton Rouge, says the story "has taken on the character of beekeeper folklore...samples taken from honey bee colonies, both feral and managed all over Louisiana, have shown no sign of African bee genotype influence." Still, some beekeepers remain suspicious.

Early Arrival of Africanized Bees in California

A widely-publicized Africanized bee arrival in the U.S. was the discovery of a colony of Africanized bees in Lost Hills, CA, in 1985. A bulldozer operator noticed bees attacking a rabbit and saw a dead fox near an underground burrow in an oil field near Bakersfield where he was working. The bees attacked his bulldozer too, though fortunately he was not stung as he was in an enclosed cab. He dumped a load of dirt on the hole but a few days later he still saw bees in the area so he reported them to the local animal control office.

Morphometric measurements revealed the bees to be a mixture of African and European bees. When the nest was dug up, remains of queen cells suggested that the colony was at least one year old, maybe two, and that it might have swarmed. That finding led to an extensive survey to determine if Africanized bees were established in the area. There was a chance the bees could have been carried to the area on pipe and oil drilling equipment moved from Venezuela.

A quarantine area that eventually would encompass almost 3,025 km^2 (1168 mi^2) was established. A large task force collected and examined bees from 400 feral and over 22,289 managed bee colonies for evidence of Africanization in the restricted area. The survey suggested that the original African genetic material was being diluted by the more common European genetic material of the surrounding managed bee colonies.

A total of 12 colonies, determined to contain some degree of Africanization, were destroyed. The cost of the effort was pegged at $1.2 million—$100,000 per Africanized colony found. The quarantine was lifted after 6 months.

The California discovery became a media event. with daily press briefings to meet the information demand of a press that took a strong interest in the discovery of killer bees in the U.S. Particularly noteworthy were the rumors surrounding the program. At one point, a radio station reported that all 22,000 colonies in the quarantine area were going to be destroyed. Arizona closed its border to all California bees. As the search and discovery program dragged on, beekeepers began to become critical of the program and concerned that they might lose their livelihood. One beekeeper, caught removing honey from his colonies, was arrested.

The California task force identified Africanized versus European bees using comb measurements (for feral colonies) and by body size (morphometric measurements). Since nobody knew what size bees were before the Lost Hills find somebody had to measure bees. Some 10 years later, additional data indicates that there are smaller-bodied "desert bees" in California. It now appears an error could have been made designating Africanized lineage in the managed colonies sampled after finding the initial feral Africanized bee colony.

Texas

As expected, Texas was the first U.S. state to experience Africanized bees. Even before Africanized bee colonization Texas authorities established a multi-agency Africanized bee advisory committee. They prepared an action plan and developed educational materials in English and Spanish for use by school children. (Figure 2–3) They also prepared a Bee Kit packaged in one of the cardboard traps used to monitor the spread of the Africanized bee.

A cooperative federal/state effort systematically gathered information on the direct effect of the Africanized honey bee on the human population in Texas. In the first 26 months, 78 stinging incidents were recorded with Africanized honey bees but only 29 with European honey bees. Africanized bee attacks resulted in higher numbers of stings and included more multiple-victim incidents compared to those involving European bees. In the second year of colonization (1992), the number of attacks attributed to Africanized honey bees increased by 11.5%.

Figure 2–3. Africanized bee alert extension material. (Courtesy Texas A&M Extension)

In May of 1991, Jesús Diez became the first person in the U.S. of record to experience a stinging incident from Africanized bees. He was mowing a lawn at a trailer court in Brownsville when bees suddenly started attacking him. He received 18 stings—when the feral colony was found and sampled, it was confirmed to be Africanized.

More stinging accidents were reported in 1992 as the Africanized bee continued northward into the larger cities of San Antonio and Corpus Christi. The first human fatality in the U.S. due to Africanized bees was July 15, 1993. Lino Lopez, an 82-year-old man, was stung at least 40 times when he tried to eliminate a colony of bees that had settled into an abandoned building near his home. The stings caused fluid to accumulate in his lungs. Analysis of the bees showed them to be Africanized.

By the end of 1995, three human fatalities had occurred in Texas from Africanized bee stings, only one a massive stinging attack. Texas Bureau of Vital Statistics records for 50 years prior to Africanized bee arrival were 0.91 deaths/year from honey bee stings. Thus, to date, the death rate from bee stings has not increased in Texas.

The Texas Apiary Inspection service of Texas A & M University continues to maintain a trap line to document Africanized bee movement and assist its quarantine program. Blue plastic cardboard (piñata) traps are fitted with an inner cardboard lining (available from Nucleus Industries) and baited with Nasonov pheromone. The inspectors ID all swarm catches using FABIS and the computerized morphometric analysis. Current distribution for Africanized honey bees in the U.S. and Texas can be accessed at websites listed at the end of this chapter.

Arizona and New Mexico

The Africanized bee arrived in New Mexico and Arizona in 1993. It is now widely distributed in the southern portions of both states. Feral populations seem especially high in the Tucson and Phoenix area of Arizona, where more than half the population of that state resides.

The first human fatality from Africanized honey bees in Arizona occurred October 6, 1995, east of Phoenix when an 88-year-old woman was stung over 1000 times. The second fatality occurred only 12 days later when a 66-year-old man was stung approximately 130 times. Four total deaths were attributable to the Africanized honey bee in the first 7 years of their being in Arizona.

In Arizona, Africanized bees swarm throughout the year with a bimodal frequency (highest numbers in spring and fall corresponding to best forage opportunities). The bees appear to thrive in the cities where water is available rather than at desert sites. Africanized bees have shown a preference for nest sites in areas of high humidity such as ground-level water meter boxes. In Tucson 23 swarms were reported in these boxes in 1993, 250 were found in 1994 and over 300 reported in the first six months of 1995. Steve Thoenes, a bee supply dealer, now bee-removal expert in Tucson, reported "swarms have increased by at least 1000% as the area [Tucson] has become Africanized."

Justin Schmidt, a bee researcher at the USDA bee lab in Tucson, believes "swarm traps, when properly deployed in areas of high risk of a bee stinging incident, are a viable and effective means of control of unwanted swarms of Africanized and feral bees." Over 10 commercial companies have begun to deploy swarm traps at sites such as golf courses, school grounds and the Desert Museum to trap swarms as a control technique. Since conservative estimates of the cost of swarm removal are $100–200, the cost of monthly trap maintenance ($7.66 plus an additional $12.64 to remove trapped swarms and provide a replacement trap) is a good preventive measure.

Gerald Loper, retired bee scientist from the USDA bee lab in Tucson, has followed honey bee nesting success in feral sites of rock cliffs at an isolated desert site outside Tucson. Varroa mites have greatly reduced colony longevity and the number of occupied sites is lower (48 in 1999) than previously (200, mostly all European honey bees, in 1995). Africanized honey bee colonies live longer than European honey bee colonies with the intense Varroa selection pressure.

Arizona and New Mexico officials decided that the Africanized honey bee is not quarantinable. The State Africanized Honey Bee Advisory Committee in Arizona elected to pursue an educational campaign. An Africanized honey bee training manual was produced along with several excellent videos on Africanized honey bees. Targeted audiences are school children, fire company personnel and pest control operators. Ordering information is at website http://ag.arizona.edu/pubs/insects/ahb/ or contact Dave Langston at the Univ. of Arizona, Tucson, AZ 85719 (Langston@ag.arizona.edu).

The Main Population Reaches California

The first California Africanized honey bee feral colony was detected on October 24, 1994. There were 9 additional swarms/feral colonies, confirmed as "highly Africanized" using morphometric and mtDNA analysis, within the next 8 months.

Surveys continue to document Africanized bee colonization northward into California. As of mid-1996, 26 Africanized bee colonies were collected, all confined to the southeastern region. In 1997 there was a spike in Africanized honey bee numbers; 7% of captured bees were Africanized in SW California—by 1998, 25% were Africanized. The population has spread northward moving into the Los Angles basin by 1999. Spread over low-elevation coastal mountains into the Central Valley has occurred but how extensively the Central Valley will be colonized is unknown.

As of late 1998 19 confirmed Africanized honey bee stinging incidents occurred involving 34 people in California. There has been only a single fatality—an 83-year-old beekeeper who bumped a hive with his lawn mower. The median number of stings of other accidental encounters has been 15 but one accident resulted in one individual receiving more than 500 stings. The California Department of Public Health, Division of Communicable Diseases, distributes incidence information and expert medical information on sting reactions.

Estimates vary on the impact the Africanized bee will have on California. Bees pollinate 47 different commodities worth $1.8 billion annually compared to the $55 million direct value of their products. Since all but the lower Imperial Valley is north of 34°N latitude and higher in elevation, geography "may provide refuge for European bees" according to Dr. Robert Page, UC Davis bee scientist. Establishment of a bee stock-center with broad-based support to breed bees that might compete with the invading population has been proposed but has yet to be realized.

Dr. Page says the Africanized bee population expansion in California most definitely is "not due to genetic mixing" based on his sampling data. He is concerned that the loss of feral colonies to mite predation favors colonization as it "will leave the currently invading Africanized honey bees with reduced competition for nest sites and resources." This could likely aid continued expansion into the dry southern interior of California and their spread in the Los Angeles basin and Central Valley.

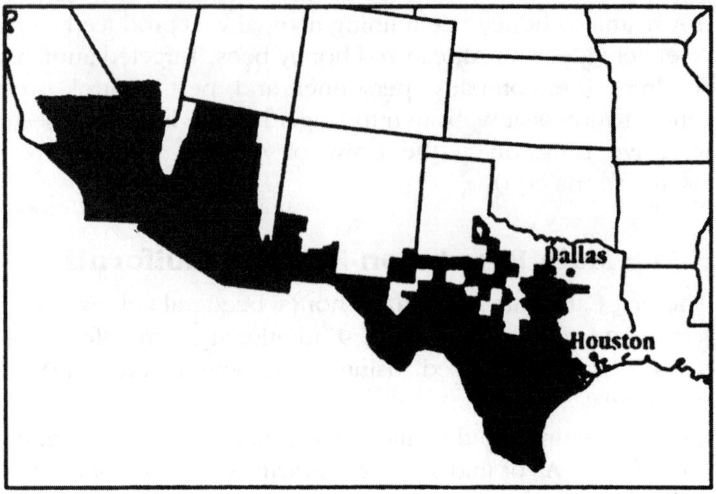

Figure 2–4. Mid-2001 distribution of Africanized bees in the U.S. (Source USDA)

Norm Gary, well-known bee scientist now retired from UC Davis, addressed the 1990 Western Apicultural Society (WAS) conference regarding hobbyists and Africanized honey bees. Since it will be impossible to maintain 100% European colonies, Dr. Gary felt that "urban hobby beekeeping should be abandoned [since] it is too risky for everyone involved" once the bee population is Africanized. Although he is confident beekeeping will survive for those who can change and adapt, he believes "hobby beekeeping may never again be quite the same."

Spread Eastward

Trap-lining has documented a slowing of Africanized bee movement northward in Texas. Explanations of why the population has still not expanded east of Houston and the bees not colonized Mississippi, Alabama or Florida to date are wanting. Each year some swarms hitch rides on trucks or railroad cars and show up in more northern locations. In July 2000 Africanized bees stung and killed a goat in Low Moor, VA, and were found clustered on the outside of an European bee hive. Other states with isolated Africanized bee finds include WV, ME, and IL.

It will be very interesting when the Africanized bee reaches Florida. Florida is a major beekeeping state and, like Texas, has a strong apiary program. Florida authorities have been proactive, seeking to prevent an accidental introduction. In addition to maintaining several hundred pheromone-baited trap hives and elimination of swarms/colonies at the docks, Florida has been sampling bee colonies located near ports-of-entry to be certain that colonies remain European. The negative impact of Africanized bees will likely be significant in Florida.

Have They Slowed Down?

Since arrival in the U.S., the Africanized bee has continued to expand its range. The good news is that the Africanized honey bee has slowed its movement rate since arrival in the U.S. The front spread 43% more slowly within the U.S., (225 km [140 miles] the first 12 months and 210 km [130 miles] the second 12 months) compared to a calculated movement rate of nearly 383 km (238 miles) through Mexico. It moved westward more quickly than northward, but reported finds may have resulted from multiple arrivals from Mexico rather than spread from within the U.S.

Evidence collected by Bill Rubink and associates at the USDA Weslaco Bee Lab suggests several factors may be involved in slow U.S. colonization. Studies at three sites (24°N latitude [Mexico], and 26° and 28° [Texas]), showed "rates of Africanization increased rapidly" after initial arrival but Africanization in Texas was "slower than in more southern populations." The researchers attribute this "to the high density of feral EHB [European honey bee] and more temperate climatic conditions...[and]...the Varroa mite may also have an impact."

Frank Eischen of the USDA Weslaco Lab believes a "Varroa-altered landscape" resulting in fewer feral European colonies favors the Africanized honey bee. Paul Jackson, head of the Texas A & M Apiary Inspection program, is certain Varroa mites are negatively impacting spread of Africanized bees. Spread of the Africanized bee population, as monitored with the bait hives, has been reduced each year. Only two additional counties were added in 1996 and a single additional one in 1997. The trap line evidence shows that European-race bee colonies, protected by beekeepers by use of Apistan, are now more frequent in bait hive captures than Africanized bees.

A large feral European bee population has often been cited as one factor that might reduce the impact of the Africanized bee in North America. It is now apparent that the feral bee population has dramatically decreased in the U.S. A Weslaco study, for example, found a 20% death rate in the feral population at the 28°N latitude study site (Welder Wildlife Refuge) during 1992–94 monitoring, but the mortality rate was 57% during the first 9 months of 1995; the difference was attributed to Varroa mite depredation of bee colonies. In Mexican studies, the feral population of Africanized bees suffered less from Varroa than feral European-race bees.

In California Rob Page of UC Davis found 75% of feral colonies in the Sacramento, CA, area dead after Varroa mites appeared. Along the California coast, where mites are uncommon, feral colony survival is significantly higher. California feral colonies were more likely to be found with mite infestations where commercial beekeepers maintained apiaries.

Paul Jackson says the Africanized honey bee has made Texas beekeepers better beekeepers because they are forced to requeen and inspect colonies more frequently. Many fewer hobby beekeepers are in the quarantine areas than before colonization. The Africanized bee has not hurt the queen rearing business or disrupted beekeeper movement of colonies for pollination or overwintering. Some 30–40,000 pollination colonies are moved between the lower Rio Grande Valley (vegetable production) to high plains (seed crops) for pollination rental. To move from a quarantine area (a county with confirmed Africanized honey bees), beekeepers must obtain a certification that confirms (after morphometric tests) that all colonies are European, not Africanized. Beekeeper regulatory compliance is very high according to Jackson.

Eventual Distribution in North America

Virtually everyone who has studied the Africanized bee believes it will establish a large feral population in the southern U.S. Dr. Orley (Chip) Taylor critically analyzed the bee's early movement history to predict where they may eventually colonize. He hypothesized different limiting climatic factors to produce a map of projected expansion into the U.S. (Figure 2–5) His use of a 240-day frost-free winter was the most conservative, a

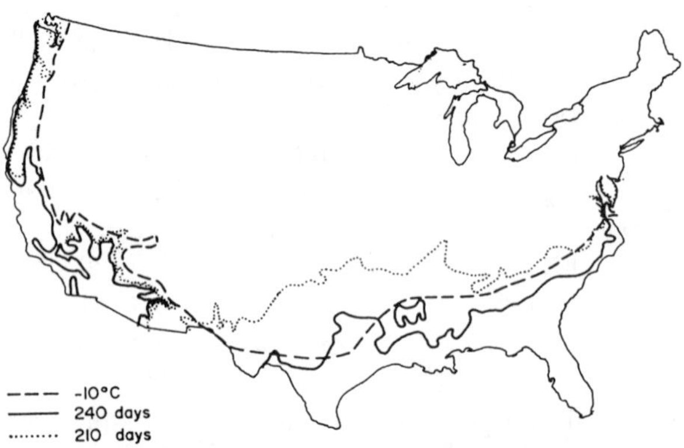

Figure 2–5. Predicted eventual distribution of Africanized honey bees in the U.S. based on 3 temperature limitations. (From: O. R. Taylor 1977 Bee World *58:19–30)*

line at -10°C (14°F), representing the lowest annual temperature was intermediate; the area of greatest predicted distribution was a line of 210 days of frost-free winter weather. In general, the predictions ranged from Africanized bees occupying limited areas in the south tto virtually all of the southern U.S. with extensions up both coasts.

Dr. Taylor, with Marla Spivak, updated these predictions in 1984, factoring in additional information that showed stabilization of the movement of the bee southward into Argentina. They predicted a 300–400 km/yr (186–250 mi/yr) spread rate in the southern U.S. over a 13–20 year period of expansion. This resulted in a slightly more conservative scenario of eventual Africanized bee establishment.

Other predictions have forecast a wider expanse of eventual spread in the U.S. Al Dietz, assessing factors such as overwintering ability and the impact man-assisted redistribution of African genes may have in the U.S., felt it possible for "Africanized bee colonies to exist in almost all of the United States and some parts of Canada." A prediction in a paper by Southwick, Roubik and Williams projected an eventual distribution of Africanized bees throughout much of the southern U.S., with more northerly extensions up both the west coast and eastern seaboard.

Taylor believed both these latter predictions to be unlikely due to their failure to consider that Africanized bees are short-lived as overwintering adults. The bees do not occur in areas where winter duration (measured by pollen/nectar forage availability) exceeds 3.5 months in either Argentina or South Africa.

Mark Winston feels no single trait will prevent Africanized bees from moving into cold regions, but "...the combined characteristics that make these bees tropical will limit their spread northward." His prediction in his book, *Killer Bees*, is a compromise of Taylor ("too conservative") and Southwick ("a bit extreme").

A significant unknown factor, perhaps ultimately the most important, impacting the eventual distribution will be the extent of human-assisted movement of populations of Africanized bees. In Peru, Robert Kent demonstrated that movement of colonies by commercial beekeepers was responsible for the Africanized bee spreading to the western side of the Andes Mountain chain.

African bee colonies carried directly from tropical areas to Poland and Germany have overwintered successfully but did not do as well as European bees in the region. In Africa, the African bee occupies higher altitudes up to 2500 m (8200 ft) in the Drakensburg Mountains. It invades higher elevation sites over 2400 m (7900 ft). but retreats during drought and colder weather.

It is thought that a transition zone will develop in the U.S. as has occurred in Argentina. Africanized bees may extend northward of an area of permanent colonization depending on seasonal weather conditions and colony movement by beekeepers. Both Mark Winston and Frank Eischen, among others, believe a transition zone will not be static nor clearly delineated. It may extend northward during the spring and summer or in years of mild winters.

Efforts at Italianization of Africanized bees in Brazil involving the production and distribution of "thousands of virgin Italian queens" was not able to halt spread of the bee, though perhaps such a program would be more effective in a transition zone as it could help tip the numbers toward European bees.

The predicted distribution estimates are primarily temperature-defined. USDA scientists Rinderer and Hellmich presented an alternative argument based on examination of the ecological factors influencing the distribution of the African bee in Africa. They state that rainfall, "more precisely variability in rainfall," rather than temperature, has been the "primary selective factor in the evolution of the African *scutellata*."

Thus the area Africanized bees will eventually inhabit in the U.S. depends on important unknown factors such as:

- How well is the present feral bee population adapted to the U.S. and will it recover its former numbers?
- Is the U.S. habitat better suited for the Africanized bee or better for the present occupant from Europe?
- How much will man enhance or retard the development of new hybrid ecotypes that will develop?

Rinderer and Hellmich believe conditions are such that "the potential to produce novel populations of bees very rapidly" exists. Hybrid populations may represent improvement or a negative with regard to defensiveness, ability to overwinter, honey production capability, and other heritable bee characteristics.

✧✧✧✧✧✧✧✧✧

The distribution of the Africanized honey bee has been reasonably well documented. What is your opinion on the eventual ultimal distribution of the Africanized bee? It would be nice if we could all vote for one distribution and declare that is what will happen. Africanized bee colonization can be considered an experiment but like all experiments, we need to wait for the results before we know which hypothesis (guess) was the correct one. Surely if it were a matter of voting, there would be little doubt as to which alternative might be chosen.

Information on movement in Texas, Arizona, and California continues to be updated with articles on the impact of Africanized honey bees in these three states. For current information check these web sites:

Texas A & M/Texas Apiary Inspection http://agnews.tamu.edu/bees/

USDA Tucson Bee Lab http://gears.tucson.ars.ag.gov

California San Diego Natural History Museum
http://www.sdnhm.org/research/entomology/ahb.html

University of California at Riverside http://bees.ucr.edu
www.cdfa.ca.gov/pests/honeybee/California.html

3

The World of Bees

Scientists are not in agreement on the number of bee species in the genus *Apis*. Four species are traditionally listed for the genus—namely *Apis mellifera*, *A. cerana* (formerly *indica*), *A. dorsata*, and *A. florea*. An additional five species probably are valid. They are *A. laboriosa* (closely related to *A. dorsata*), *A. andreniformis* (related to *A. florea*) and finally *A. koschevnikovi*, *A. nigrocentris* and *A. nuluensis*, all related to *A. cerana*.

Thus, the insect humans call "honey bee" may be one of several species. In the western world, a honey bee is the species *Apis mellifera*, originally described by the Swedish biologist, Carl von Linné (Linnaeus). This is so because in Europe and the Mediterranean, then the known world for Linnaeus, there is only this single species. The remaining Apis species, all native to Asia, are also commonly called honey bee, or its equivalent, by the peoples native to those areas.

Of the four traditional and better-known species of honey bees, two build a nest of parallel beeswax combs in a cavity of suitable volume. The nest may sometimes be expanded or even built outside a cavity. These two species are sometimes designated as the eastern or Indian honey bee (*A. cerana*) and western or European honey bee (*A. mellifera*) to help distinguish between them.

The smaller-bodied *Apis cerana* is distributed from the Middle East to Asia occurring in temperate regions of northern China and colder climates of Japan, as well as in the Asian tropics from India to China and the Philippines. The three related species have limited distributions on some of the Asian islands. Both *A. cerana* and *A. mellifera* live in climates as diverse as hot, steamy tropics to temperate climates with cold winters and cool summers.

Apis mellifera, the western or European honey bee as recognized by Linnaeus, occurs naturally from southern Scandinavia to the tip of South Africa and from western Africa (Senegal) to Iran and the Ural Mountains in middle Asia. Colonies live from sea level to 1000 m (3280 ft) elevation in the European Alps but up to 3100 m (10,170 ft) on Mt. Kilimanjaro in Africa. They are scare in the Sahara Desert region but present in hot, dry Oman. They also do well in the wet, humid African tropics, demonstrating the extensive variation that occurs in the species.

Subspecies or Races of Bees

With a familiar, widespread insect like the honey bee, scientists once believed it important to designate subspecies or races to differentiate between populations. This is no longer in vogue as the ecological concept of species has changed to encompass interbreeding populations possessing local variation.

The word "race" in honey bees is not the same as race (or breed) in cattle, dogs or other animals. Honey bee races are ecotypes with distinct, recognizable variation developed via natural selection due to geographical isolation. They are not from planned interbreeding as we see in dog breeds or farm animals.

What is important in the biological sense is a species. A species is a population with a shared gene pool that does not interbreed with another population. That means *A. mellifera* does not interbreed with *A. cerana* or other species of *Apis* nor, of course, with other species of animals. Interbreeding can occur among races since they are the same species. Races represent local variants in the larger overall population.

In nature, one population may be separated from another by a geographical barrier no larger than the next tree, or a road through a forest and yet the Atlantic Ocean might not be a geographic isolating barrier for another organism. Races, as designated in honey bees (including their designated subspecies scientific names), are important however in the story of the Africanized bee and, old concept or not, are still used.

Honey bee races are generally uniform where they occur, except at the interface of one population with another, where mixtures of the two may occur. Races are most frequently named, both commonly and in their latinized subspecies name, for the geographical area where they were found. Differences between honey bee races are mostly described in morphological characteristics of body size, hairiness, body color, wing venation, tongue (proboscis) length, number of hamuli (hooks) holding wings together in flight, width of hind leg metatarsus (the comb-containing portion) and other body parts.

Behavior, physiology, nest and comb characteristics and the newer biochemical techniques can and are increasingly being used to validate species and, to a lesser degree, races. The morphological characteristics used to separate species are simply outward manifestations of genes or genetic material contained in the DNA of the cell nucleus.

A striking aspect of honey bee races is the variability within, as well as between races. Beekeepers have taken advantage of such variability to select economically desirable characteristics, seeking better bees for their environment and nectar plant conditions. Rather than developing different breeds, as in cattle, bee races have been carried from one region to another, often without regard for the consequences.

Although there is not total agreement, some 24 subspecies or races, are recognized in *A. mellifera*. Prior to widespread global movement of man and animals, races could be conviently subdivided into African bee races (6 to 8), 7 Near East races and 9 Mediterranean and European races.

The distribution of 17 major races of *A. mellifera* are shown in the map from *The Hive and the Honey Bee* (Figure 3–1), including all of those from Africa. Missing are Macedonian and Greek races of the Adriatic peninsula, island races of Cyprus and Sicily in the Mediterranean, and four races farther east in the Middle East.

Figure 3–1. Map of distribution of African and European bee races. (Source: The Hive and the Honey Bee)

As many as 16 African bee races have been named south of the Sahara desert, with additional races along the Mediterranean coast. Using morphometric measurements and more recently biochemical differences, the races below the Sahara Desert are readily identifiable as different from the bee races of the Mediterranean and European region, suggesting that a long geographical isolation has existed to separate them.

The bee races of Northern Africa, along the Mediterranean, were briefly imported into the United States in the late 1800s. The *American Bee Journal* advertisement Figure 3–2 is for the punic bee of Northern Africa. They were not found to be commercially useful and importation did not continue. We know too little about biological differences between these and the other African bee races.

The several geographical races of Africa below the Sahara desert show considerable variation. There is a tremendous variation in climate, rainfall and environmental conditions on the African subcontinent so such variation in honey bees is not surprising. The largest- and the smallest-bodied bee, the darkest- and lightest-colored bees, and the most and least defensive of all bee races exist in Africa. The African bee races likewise vary in ease of manipulation and honey production capabilities.

While searching for a bee to carry to Brazil in 1956, Professor Kerr believed that there were only three subspecies of honey bees present in Africa. According to rules of zoological nomenclature, the name of the most widespread race was *Apis mellifera adansonii.* Therefore this was the name Kerr applied to the bee he subsequently carried to Brazil.

Figure 3–2. Early American Bee Journal advertisement for punic bee (from northern Africa).

Professor Ruttner, an eminent European bee scientist, reexamined the issue of African bee races in 1975. He discovered that the widespread African bee was different from *adansonii* as it was described and named to science. Ruttner reported *A. m. adansonii* to only occur in western Africa from Senegal to the jungle around the Zaire River. The subspecies most widely distributed in Africa was, according to him, *Apis mellifera scutellata.*

In his revision of the African bee races, Ruttner accepted two "varieties" (*A. m. litorea* and *A. m. monticola*, both existing within the limits of the more widely distributed eastern African bee) as separate races along with subspecies designation of *A. m. unicolor* for the bee inhabiting the islands of Madagascar and *A. m. capensis* for a race at the tip of South Africa. His classification thus totaled six sub-Saharan subspecies (populations) of honey bees.

Other scientists gradually accepted the conclusions of Ruttner and his interpretation of African bee races. Therefore the name of the bee carried by Kerr to South America has changed in the literature from *adansonii* to the new designation of *scutellata*. It is the same bee population only with a different name.

Apis mellifera scutellata

The Latinized subspecies name *scutellata* was selected because the bees possess a distinctive yellow scutellum (part of the thorax). This bee is small with few body hairs, has one or more yellow color bands on the abdomen, and a distinctive wing venation. Drones are dark-colored.

In nature, *A. m. scutellata* occurs from the Sudan to South Africa, principally in eastern Africa. The climate of this area, predominately savannah or open woodland, is hot, with long dry seasons but with many abundant honey plants. Guy, a South African beekeeping authority, called this region of Africa "one of the most favorable beekeeping areas in the world." As much as one-half of the beeswax in world trade originates from beekeepers and honey hunters in this region of Africa.

A. m. scutellata is well-adapted for its habitat. Because the climate is dry, both swarms and colonies survive outside in unsheltered sites, exposed to the elements. Common with tropical bees, *scutellata* is migratory both in swarming and in absconding. Even small swarms have a good chance to survive, especially as they tend to migrate to areas with better food resources, foraging as they move. Once the homesite is accepted, the population builds up rapidly under favorable conditions and colonies quickly again begin swarm preparations.

A. m. scutellata survives in climatic areas of both low and high temperatures over much of its natural range in Africa. Colonies occur at 2000 m (6562 ft) altitude where snow persists on the ground and minimum temperatures of less than 0°C (32°F) are common for six months of the year.

The bee has acknowledged good and bad qualities. In Africa *scutellata* had long been recognized as "bad tempered" or aggressive by beekeepers from South Africa to Ethiopia. A noted authority on beekeeping in Africa, F. G. Smith, characterized the bee in Tanganyika (present day Tanzania) thusly:

> "Viciousness is the first characteristic which beekeepers and others encounter.... The colonies always appear to be alerted, ever ready to defend the hive, and on occasion the whole colony goes berserk and stings every living thing in sight. A beekeeper can usually smoke a colony and get it under control while he collects the crop. But if he appears near the hive a second time within several weeks the bees are on him."

The same bee was also recognized as highly productive, certainly more profitable than European bees which have been repeatedly introduced to Africa without much success. *A. m. scutellata* was the world record holder in honey harvested from a single colony in one year (256 kg – 565 lb) at one fixed site. Smith called the bee "the best of all the honey producers in the tropics" in a 1953 *Bee World* article. He recommended that it be considered for importation to India or to the Far East tropics.

Figure 3–3. Native depiction of nighttime harvest of bee colony in a tree.

Both honey production and the bees' defensive reputation were factors considered in the decision to import the bee to Brazil. Within 10 years post-importation, Nogueira-Neto, a Brazilian bee scientist, seeing firsthand the rapid progress made by the bee in Brazil, wrote "we would urge that the importation of any *adansonii* stock, or their hybrids, should be forbidden by any country not yet invaded." Unfortunately, this advice was too late for the Americas!

Other African Races

Besides *scutellata*, another well-known African bee is the cape bee, *A. mellifera capensis*. It lives in the extreme southern portion at the tip of South Africa. This bee race has attracted interest because its workers have a large number of ovarioles and worker ovaries develop more rapidly than in other bees in the absence of the queen. Workers cannot mate but a high percentage of them parthenogenetically produce diploid eggs that develop into new queens (thelytoky) when the colony loses a queen. More recently this same development has been determined as occurring in *scutellata*. This developmental difference has been suggested as an adaptation to harsh environments but why these two bee races and not others regularly replace queens this way is unknown. Certainly other bees live in equally difficult environments.

Within the range of *scutellata*, two other bee races, originally called varieties, are recognized. *A. m. monticola* is a black bee found on several East African mountains of Ethiopia, Kenya and Tanzania, including the famous Mt. Kilimanjaro, between 1829 and 3658 m (6000 and 12,000 ft) elevation. It is a gentle bee according to Smith and others. Hybrids between it and *scutellata* exist in transition zones where the distribution of the two bees meet. At higher elevations *monticola* maintains a separate population.

Why Introduction?

Why the introduction of African bees into Brazil was even considered in the first place is a moot question now in view of what happened. The intention of the program was not to simply release the bee but rather to interbreed the introduced African bee genetic material with European bees as a means of developing a bee better suited to tropical conditions.

Dr. Kerr had the support and financial backing of the government and the beekeepers of Brazil in his quest to find a bee better suited to tropical conditions. The defensiveness (bad temper) of the bee in Africa was well known but it was also known to be "very prolific, better than the Italian bee in honey production, colony development and adaptability to climatic conditions." It was the world-record holder for honey production and had even been recommended as a bee suitable for introduction into other tropical areas.

In the early 1950s, Brazil, despite being a large country with widely varying flora and a climate appropriate for bees, was only 27th in worldwide production of honey (approximately 6,500 tons/year). Beekeeping was a growing, thriving industry in the south of Brazil. Just as North American settlers carried bees to western lands, settlers were taking bees to the tropical north of Brazil, long considered the frontier of the country but European bees did not do well there. It is understandable that beekeepers and bee scientists would look to introduction of bees from elsewhere to advance honey production.

Ironically, the original intention of the importation has been achieved—a bee suitable for the American tropics now has been produced! There is general accord that there were essentially no feral colonies of honey bees in South America prior to African bee introduction despite 200 years of beekeeping with European bees. Estimates are that the offspring of Africanized bees have increased to between 50 and 100 million feral colonies in the Americas. Tropical bee specialist Dave Roubik, of the Smithsonian Tropical Research Institute in Panama, estimates one trillion (10^{12}) honey bees now inhabit South and Central America.

Unfortunately the successful introduction comes with a down side—the bee is not readily manageable and is overly defensive!

Previous Introductions of Bees

Since honey bees are not native to Brazil, or any area of the Americas, humans carried bee colonies to the Americas as they settled the continent. The first honey bees introduced into the Americas were European bees. *Apis mellifera mellifera*, the German bee from northern Europe, is believed to be the first race introduced. It came on ships crossing the Atlantic during the winter months before 1622. The German bee initially spread rapidly, perhaps even without human assistance. It was the only bee race present for about 200 years.

The German bee is defensive and susceptible to both brood disease and wax moth depredation. so it was rapidly replaced as the preferred bee for beekeeping in the late 1800s when queens of *A. mellifera ligustica*, the Italian honey bee, became available. Of the first 14 colonies of Italian bees that arrived in 1859, only six survived the first winter. These were used by their owner, Phineas Mahan of Philadelphia, to rear queens to sell the following summer. As in Europe, U.S. beekeepers quickly replaced German for Italian bees.

In addition to Italian bees, at least seven additional subspecies were imported into the U.S. between 1859 and 1922. One, *A. mellifera carnica* (the Carniolan bee), originally a bee of very limited distribution in Austria was distributed by beekeepers to Eastern Europe, Austria, Germany and into England 100 years ago. Charles Dadant brought it to North America.

The Carniolan bee has two distinctive color phases. It is best known as a bee that is black in color with no yellow. This color phase is sometimes called the dark version of an Italian bee, to which it is most closely related. Carniolan bees overwinter better on fewer stores, are more responsive to environmental conditions than Italian bees and their colonies build rapidly in the spring. Unfortunately it also swarms more often. The Italian and Carniolan bee were the two most widely distributed bee races in the world prior to the introduction of African bees into Brazil.

Another European bee race introduced into the Americas was the Caucasian bee, *A. mellifera caucasica*. Originally the bee lived in valleys of the central Caucasus Mountains. Beekeepers in eastern Europe and the former USSR preferred this bee over the German bee so the stock was distributed more widely. It never became popular in the Americas as it is susceptible to Nosema disease, making it a poor overwintering bee, and it uses large amounts of propolis. Recently there has been renewed interest in Caucasian bees as they display more resistance to mites than Italians.

The bee of Spain, the Iberian bee, was undoubtedly carried to Spanish-settled areas of South and Central America. The bee population in Spain has two distinct mtDNA types. One influence is from *A. m. mellifera*, the German bee, and the other is from *A. m. intermissa*, the bee of northern Africa above the Sahara. Fortunately the northern African mtDNA is easily distinguished from the mtDNA of *scutellata*. (Chapter 4)

Fearful of what was causing heavy colony losses in England U.S. authorities enacted a 1922 Federal law banning the importation of bees into the U.S. No such restriction covered South or Central America and bees have continued to be carried into the Americas from Europe and Asia with different waves of immigrants. For the last 100 years, Italian and German bees have been the bee races of choice, at least for southern Brazilian beekeepers.

Why Did the African Bee Succeed?

It seems appropriate to ask why the African bee succeeded in the American tropics whereas the European bee did not. Probably no one factor can be listed as the reason but several factors that have been suggested are as follows:

❖*Usurpation (queen parasitism) of European queens by Africanized swarms:*

Female social reproductive parasitism, or usurpation, is one of the more controversial aspects of Africanization of managed European bee colonies. The NAS committee in 1972 learned about it from beekeepers in Brazil. There are a number of anecdotal reports of swarms or a small cluster of worker Africanized bees moving into existing colonies of bees and a docile colony almost immediately changing into one that is defensive. European genetic material is replaced (usurped or parasitized) by Africanized bees but the reverse does not seem to occur.

Scientists of the USDA lab in Baton Rouge found queen parasitism in queenless and weak/stressed colonies in Venezuela and again in studies conducted in the Yucatan of Mexico. Al Dietz and students, in studies in Mexico, calculated an usurpation rate of almost 10%, of which approximately 40% were of strong colonies. Usurpation of European queens was "significantly higher and more successful" than the reverse. Usurpation is also reported as common in small queen mating nucs.

❖*Parasitism of European colonies by Africanized drones (drone parasitism):*

Drone parasitism is a well-documented behavior in Africanized bees. Africanized drone bees drift into European colonies and are readily accepted, but the opposite is not true. Africanized bees rarely allow drones of any racial mixture to enter.

The net result of drone parasitism is an increase in Africanized drones and a drop in European drones in an area. Africanized colonies raise more drones to make up for drones lost to drifting; European colonies raise fewer of their own drones following an influx of Africanized drones. Africanized colonies also raise proportionally more drones than do European colonies and raise them early in their population build-up cycle.

❖*Mating differences between tropical and temperate bees:*

It was intuitively assumed that Africanized bees have a mating advantage. However Mark Winston and Orley Taylor agree that experiments to demonstrate something unique or different in Africanized bee mating biology have not been successful. The sex attraction pheromone and mating behaviors appear to be the same in both bees. Africanized bees do everything a little faster it seems, including eliminate potential rivals sooner, fly earlier and begin egg-laying more rapidly than their European counterparts.

Tropical bees mate slightly later in the day, a likely adaptation to high midday temperatures. Temperate-zone bees moved to the tropics do not shift their mating time to later in the day, suggesting this behavior is genetically programmed. No space separation occurs in nature as both Africanized and European bees mate in the same Drone Congregation Areas.

Africanized queens mate later in the day than European queens. Africanized drones however fly early as well as later in the day. Africanized queens and drones fly shorter distances for mating and they become concentrated in numbers, especially useful traits when combined with high feral populations. In areas of high European drone populations, Dr. Taylor found African queens rarely mated with European drones (0–6%) while European queens mated more frequently with European drones (20–35%) suggesting assortative mating occurring in African bees.

❖*The "numbers game":*

Africanized bees are able to prosper because they reproduce sooner and more often. They swarm an average of 3–4 times per year, and swarms have a high probability of survival. One early study by Gard Otis calculated as high as a 16-fold reproductive advantage in favor of Africanized bees in French Guiana. Reproductive rates are lower in Mexico, Costa Rica and Panama but are still higher than we know for European bees.

Africanized bees rear proportionately more drones than European bees and drones are present in Africanized bee colonies longer during the season. Africanized bee colonies also generally have more drones at any one time than European colonies.

Professor Ruttner of Germany, drawing on his experiences in Africa, believes a "simple genetic consideration" can explain the numbers advantage for Africanized bees. He reasons that Africanized bees have many swarms and most survive at least to reproduce drones and swarm again so the selective value of Africanized bees compared to European bees "easily may be 10:1 or even more." He considers the European bee "permanently outnumbered to an extreme degree."

If we accept that drones fly 7 km (4.3 miles) and conservatively estimate a density of 6 colonies/km^2 (16/mi^2),though numbers higher than that have been found, the drones in an apiary of a beekeeper would compete against the drones of 3695 feral Africanized honey bee colonies for the opportunity to mate with a virgin queen according to calculations of

David Roubik, of the Smithsonian Tropical Institute. Africanized bees excel at the "numbers game"!

❖*Ecological factors favor initial colony success in Africanized bees:*

The Africanized bee has some special behaviors as a colonizing or pioneer species that favor survival. These include:

- adaptability to utilize a wide range of nest cavities, as well as ability to successfully nest outside if cavities are sparse;
- dispersal ability permitting it to traverse unfavorable terrain (migrate) in search of a more favorable site;
- absconding behaviors allowing the leaving of sites with few resources, areas with heavy predation, or locations otherwise unsuitable to reestablish at locations that are more favorable;
- swarm amalgamation facilitating survival of even small swarms;
- superior competitive foraging behavior emphasizing pollen-collecting behavior; and
- the ability to increase metabolic activity under colder temperature conditions.

Collectively considered, the Africanized bee has succeeded in the American tropics because it is better-suited for the environment. It has mechanisms to insure it maintains its own genetic identity while able to change other races over to its characteristics, thus ensuring survival and population growth.

✧✧✧✧✧✧✧✧✧

The phenomena of Africanization has been spectacular if nothing else. The bee is a tropical ecotype better suited biologically for tropical areas than European bees. David Roubik perhaps expresses it best in calling the African bee introduction to the Americas "a vast experiment...in some ways...comparable to the abrupt removal of much tropical forest during the past decades, since it, too, potentially, affects many thousands of species." As tropical forested areas shrink, so does biodiversity of plant and animal life. Africanized bee colonization may be having the same effect on bee diversity.

Humans seem to still have little appreciation of how loss of biological diversity, through destruction of tropical forests or introductions of foreign species, may affect their own fitness in the long run. Africanization of the Americas is yet another lesson of negative consequences of exotic species introduction from which we all must suffer the consequences. Can we learn from our past mistakes or must we continue to repeat them time and again?

4

What's in a Name?

What exactly is the Africanized bee? The answer, in part, depends on how precisely you want to answer the question.

The bee introduced into the Americas has been called many names. Initially it was called the African subspecies (race)[1] *Apis mellifera adansonii,* now *A. m. scutellata*. Popular names have included African, Brazilian (even latinized to *A. m. brasiliensis*) and, commonly since 1972, Africanized bee. The press name "killer bee" is highly objectionable to beekeepers, though the general public immediately recognizes this name.

A few scientists prefer return to the common name African or African Bee in the Americas. Orley Taylor, unarguably the best informed and most knowledgeable scientist who has studied the bee, uses Africanized bee but prefers neotropical African bee. In a 1991 review book, the editors Marla Spivak, David Fletcher and Michael Breed titled it the "African" bee but the authors of 18 chapters used the name Africanized.

The name game revolves around what the bee is genetically. The most widely used name, Africanized bee, implies that the bee is a hybrid. Is that the case? With the present state of knowledge the answer is both yes and—is the bee a hybrid?—no!

Morphometric Analysis

Traditionally, morphological body features, increasingly involving many precise measurements of anatomical features, have been used to identify species of insects. Morphological measurements were the basis of differentiating between the races of honey bees and have been used by European scientists in an extensive search for honey bees with longer tongues. Morphology is a reliable indicator because the African bee is about 10% smaller than the European bee. (Unfortunately it is not possible to tell European from Africanized bee just by casually looking at the bees.)

Professor Kerr, the scientist responsible for bringing African bees to Brazil, was the first to use morphometric measurements to differentiate

---

[1]There is no general agreement on correct use of subspecies or race for a geographical variant of a species. The modern concept of a species as a population of interbreeding individuals implies that variation is normal. I use race and subspecies interchangeably here and will use the common, not latinized name, as a label.

African from European bees. Howell Daly, a taxonomic specialist at the University of California, Berkeley, expanded the analysis used by Kerr into a multivariate statistical analysis using 25 wing, wing vein, hind leg, and sternum (lower abdominal plate) measurements along with number of hamuli (hooks) of the adult worker wing.

The expanded morphometric measurement technique was tested on a large sample of bees and was found reliable to identify Africanized bees within 0.5% for samples of bees from a hive and to within 95% for an individual bee. That is, if you took 10 bees from one colony and did all 25 measurements, you could correctly ID the colony as European or African 99 out of 100 times; if you had only one bee, you would be correct 95 out of 100 times.

Using an extensive series of measurements, Dr. Daly found that Africanized bees in South America differed slightly from African bees sampled from Africa. He concluded "they are closer morphometrically to African bees than to European bees in North America." A three-dimensional diagram of the relationship looks like the graph below. (Figure 4–1)

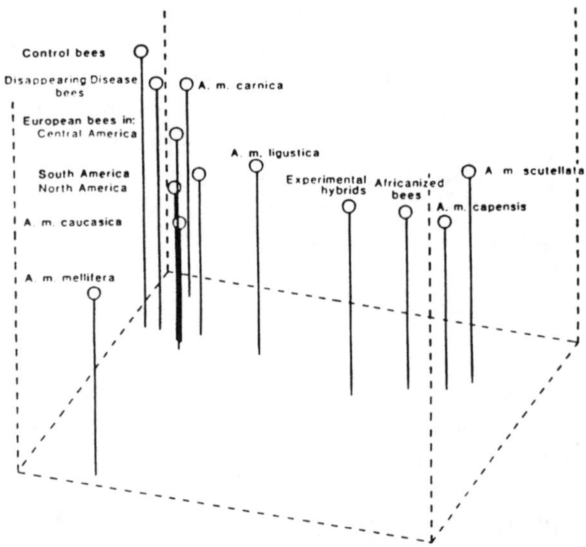

Figure 4–1. Bee race relationships from morphometric measurements. Note African races grouped at right with hybrids in middle. (From: Daly 1991 The "African" Honey Bee)

Dr. Daly, with co-worker Balling, using morphometrical analysis of adult worker bees, determined that bees invading Panama in 1982 were "almost indistinguishable" from the feral bees they examined in the early 1970s from Brazil. Retired California bee scientist, Norm Gary, and co-workers concluded the same thing when they looked at the bees invading Honduras in 1985. Orley Taylor at the University of Kansas states that

"African bees" that arrived in the U.S. in 1990 were "genetically...little changed from those in South Africa."

Validation of the accuracy of using morphometric measurements to ID bees has been confirmed experimentally. It still remains possible to tell European from African bees even when European bees are raised in Africanized colonies and vice versa. Morphological differences remain when bees are starved or raised under nutritional stress, when workers are raised in drone cells, and even when Varroa-parasitized bees are compared to healthy bees.

It takes two to five hours for a complete morphometric sample of 10 bees. You need a good microscope, appropriate computer software, a digitizer (to take accurate length and angle measurements), and skill in slide making.

A more rapid ID system, termed FABIS (Fast Africanized Bee Identification System), was developed by the USDA Lab in Baton Rouge, LA. (Figure 4–2) Accuracy is at the 90% level. The USDA trained a number of state apiary inspectors how to use FABIS and has offered continuing assistance. You need ten bees, 35-mm slide mounts, cover glass and a 35-mm slide projector. For samples that are not identifiable with FABIS, a larger set of measurements should then be performed (USDA-ID).

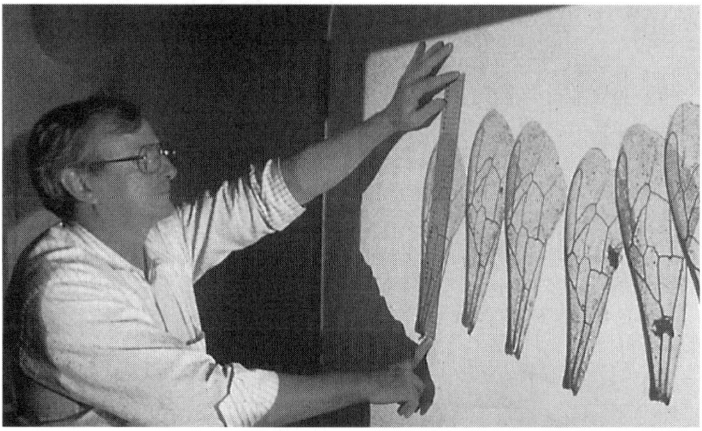

Figure 4–2. Tom Rinderer, USDA, measuring projected bee wings for FABIS identification. (USDA photo)

Daly did not feel morphometric measures alone were sufficient to properly ID African bees for regulatory activities or explain what had occurred in the Americas. Recently, Guzman-Novoa, a Mexican bee scientist who was studying for his PhD at UC Davis, demonstrated that bees less than 50% Africanized were not accurately identified using morphometrics. Comparison of results from two federal labs in Mexico showed that measurements "by different people may yield different results."

The reason Africanized bees could be slightly different morphometrically from their counterparts in Africa might be due to the small initial founding population carried by Kerr to Brazil (a founding effect), a random variation in genetic makeup of bees carried to Brazil (genetic drift, a well-documented phenomena in small populations on islands), the individual performing the measurements, or be evidence of hybridization.

Protein Analysis

Increasingly, we have developed the ability to use extremely sensitive (and expensive) biochemical methods to tell organisms apart. One technique uses allozymes (protein) to differentiate European and African bees. This analysis, (electrophoresis), utilizes different migration rates of proteins dissolved in a liquid or gas, after an electrical current is applied. Specific proteins can be accurately localized despite the existence of thousands of proteins within an individual bee. The technique must be done in a suitably equipped laboratory. It is slow and expensive but offers the prospect of accurate ID of bee origin.

Electrophoresis can separate the hydrocarbons of the insect exoskeleton. Biochemist Carlson was one of the earliest to report the ability to distinguish drones of African, European and the Africanized bee in South America using this technique. Following analysis of wax and cuticular chemistry, he concluded there was little change in the African population in the Americas from the African source population.

Brazilian scientist Lobo and associates analyzed 11 bee samples collected from tropical to subtropical climatic regions across Brazil and Uruguay. They found both morphometric and protein differences in the bees. Another Brazilian, Buco, and his colleagues sampled bees from southern Brazil to northern Argentina for allozyme frequency. They found no European-race proteins in Brazil but more were present as they sampled bees farther southward.

DNA Analysis

Analysis of proteins, of course, can be combined with morphometric measurements. Increasingly, the powerful technique of DNA analysis is also being used. Unfortunately some overlap, and conflict in data may occur when the different techniques are used, even on the same data.

Differentiating African from European includes examining both mitochondrial (mtDNA) and nuclear DNA of individual bees. DNA, or deoxyribose nucleic acid, is what codes genes. It consists of four proteins which sequence linearly into the infinite variety that we know as life. There are two types of DNA—the DNA of the nucleus of the cells, now being used in the courts to demonstrate probability of guilt or to set innocent persons free, and mitochondrial DNA (mtDNA), which is only found in the cell cytoplasm. mtDNA encodes for energy processes.

Rob Page, a geneticist who specializes in honey bee studies at UC Davis, has cautioned that the relatively small feral population of European bees in the Americas at the time of introduction of African bees might help explain the overwhelming amounts of African-derived DNA (and mtDNA) reported in feral bee sampling studies. Steve Sheppard, now at Washington State University, states that DNA studies have to be done very carefully. He has demonstrated that DNA can be "chopped" in different ways so apparent sub-Saharan DNA comes out as a higher percentage.

mtDNA Analyses

Only nucleic DNA of the male sperm fuses with female DNA of the egg—no mtDNA in the male sperm passes to the egg. All mtDNA in an organism therefore is only inherited from the female. Thus mtDNA is very conservative in evolutionary change. It represents a powerful method to determine maternal descent.

Initial application of DNA analysis showed "virtually no European-type mtDNA" in bees from Argentina to Mexico. A comprehensive study of Hall and Muralidharan of bee samples from 10 feral colonies from Venezuela, 19 feral swarm samples from Mexico and 3 feral swarm samples from Costa Rica, all demonstrated African mtDNA exclusively. In another study, Smith and associates reported that 9 of 10 samples from Brazil, 12 of 12 samples from Venezuela and 38 of 39 Mexican samples all had exclusively African mtDNA.

DNA Analysis

DNA analysis of bees from Venezuela by Glenn Hall showed "virtually no European DNA" patterns in the bees living in an area inhabited by Africanized bees for 20 years. In Mexico, "extremely low frequencies of European DNA" occurred in bees analyzed from an area where a feral population of Africanized bees had existed for only 15 months.

An analysis of bees in Brazil taken from the region where African bee genetic material was introduced, using all three techniques (morphology, allozymes and mtDNA), found populations "comparatively stable" in their genetic composition. Morphological and mtDNA data revealed "strong influence of the African race" but that "some European polymorphism does exist" in the Africanized population.

An extensive examination of managed bees in the Yucatan region of Mexico showed some colonies with European morphology and African mtDNA, or the reverse. When Steve Sheppard and colleagues examined both morphometric features and mtDNA of bees from Argentina, where Africanized and European bees have maintained an equilibrium or transition zone for a number of years, more than one-quarter of the sampled colonies were a mixture of European and African features. The study concluded that "Africanized bees in Argentina hybridize extensively with European bees."

Thus studies of both mtDNA and DNA independently demonstrate that gene flow is apparently consistently from European bee genetic material towards African in areas where Africanized bees invade. The process of Africanization is very rapid!

So the answer is...

In the book *The "African" Honey Bee* some of the same data was used by different contributors to support opposing conclusions. Biochemist Hall feels "these studies (of DNA) disproved the assumption that the primary mechanism of African bee spread has been through paternal introgression into extant European colonies...the European maternal lines do not become part of the migrating African feral population and...disappear through attrition." Dave Fletcher concurs and adds "hybrids between *scutellata* and European bees do not persist, but are replaced over time by essentially pure *scutellata*."

Dr. Taylor of the University of Kansas says a number of studies "...show that there are a number of barriers to gene flow from European bees into the African bee population but the converse is not true...survival of the hybrid bees evidently leads to a rapid elimination of European bees and their genetic traits." Hall attributes hybrid deficiency to better adaptation of African bee metabolism to tropical conditions with faster development time, shorter life spans and more rapid queen development time.

DNA evidence strongly indicates the bee is not a hybrid, although there are some measurable differences in anatomy. Mark Winston in his *Killer Bees* book concluded that "any subtle differences between the African and Latin American bees has come primarily from natural selection or from genetic drift away from the original African type, not from extensive hybridization with European bees."

USDA bee geneticist Tom Rinderer and co-worker Rick Hellmich on the other hand concluded that "collective results showed decisively that hybridization had taken place." Both felt there was definite evidence of "interbreeding." Colonies in the Yucatan study that had recently been requeened showed the least evidence of African ancestry. They believe "the feral bees of South America constitute a population which is genetically quite different from the parental population.... Africanized bees clearly show the influence of their European and their African parentage."

In a Panama study, Boreham and Roubik found that the Africanized bee population became more African in morphometric measurements with continued residence in Panama. Evidence suggested that pioneer swarms may crossbreed and are a different genetic mix than the population that eventually inhabits an area. For colonizing bees, evolutionary success may be very different compared to bees that might inhabit an area after initial invasion. Within the total bee population of the Americas there should be selection pressure for somewhat differing ecotypes.

The beekeepers of Brazil believe that today they have a different bee than they initially faced. (Chapter 10) The major changes are better honey production and less defensiveness. African bees in Africa have not been subjected to the same selective forces. As documented by David Fletcher, the African bee in Africa has a wider variation than could possibly have been included in the small population carried to Brazil. With time and geographic isolation, beekeeper and habitat selection, and a smaller gene pool from a small founding population, we should find some differences.

And so what are we to call this bee? As we have come to look at their behavior, delve into the proteins of bee body and cuticular wax, examine the structurre of mtDNA and DNA, explore ecology and ecotypes, and continue to take many body structure measurements, the bee is understood to be different from European bees, maybe slightly different from African bees, but not exactly a hybrid between the two.

How much difference is enough to call the bee population in the Americas a hybrid, an American ecotype, or a derived population most similar to its African inheritance? Opinions vary!

Identification in the Field

Knowing what bee we are talking about is critical to describing what has happened and what might be a solution. So how do we tell bees apart? Is looking at bee behavior in the field a reliable means of telling the difference between Africanized and European bees?

Figure 4–3. Defensiveness in Africanized bees. Note "cloud" of bees around beekeepers' heads.

Morphometric and biochemical techniques require a laboratory, computers and expensive equipment. In the field however, differences between European and Africanized bees can be very evident. Anita Collins and USDA colleagues, studying a population undergoing hybridization in Venezuela, found "defensive behavior alone not an adequate indicator for identification or certification programmes." Behavioral characteristics, whether accurate or not, however are used to help differentiate between Africanized and European bees. (Figures 4–3 and 4–4)

Most beekeepers, and the general public, believe heightened defensiveness to be the best indication of Africanized bee colonization. The contrast between European and Africanized bees is often readily apparent and many have commented on the differences. For example, Orley Taylor, with co-author Williamson, stated that dead bees are difficult to ID due to slight morphological differences but living colonies "are accompanied by distinctive behavioral traits."

Marla Spivak, University of Minnesota bee scientist, in her studies in Costa Rica (Chapter 7) used colony behavior to distinguish between European and Africanized colonies. She says "the only reasonable option for beekeepers and queen breeders is to select for and propagate the most gentle and productive Africanized colonies based on their behavior." Her field diagnosis had only a 5% error rate in assigning colonies to treatment groups when conditions did not permit morphometric or biochemical determinations before the experiments were started

Field characteristics that help differentiate between Africanized and European bees are as follows:

Morphometric Differences

Several measurements of body parts and even beeswax cell size can be used. The most reliable are:

- **Worker size**: African bees are approximately 10% smaller:

  AHB 12.73 mm vs. EHB 13.89 mm

- **Worker cell size**: Africanized bees construct slightly smaller feral-nest beeswax comb cells:

  AHB 4.6–5.0 mm vs. EHB 5.2–5.4 mm

  The recommendation is to measure 10 adjacent cells three times—if the average is 49 mm or less it indicates Africanized and if 52 mm or more it is European (no determination is recommended for 50 & 51 mm).

- **Wing beat frequency**: Africanized bees are a bit faster:

  AHB 290 Hz vs. EHB 243–270 Hz (a few at 290 Hz)

  Due to overlap, only free-flying, undisturbed bees should be recorded.

- **Fresh weight of 30 bees**: Africanized bees weigh about ⅓ less at birth when collected from the feral nest.

Ranges for 30 bees are:

AHB < 2 g vs. EHB > 2.5 g

Queens weigh slightly less (5%) in Africanized bees but ranges overlap too extensively to permit telling them apart.

Behavior Differences

There are many differences but they are only indicators, not hard and fast distinctions. It is not possible to assign a degree of probability for most behaviors and, even if many are included, accuracy that the bee is one population or the other is not improved. Great variation is possible between colonies and within a colony on different days. Some of the reproductive behaviors with more consistent differences are:

- **Swarming behavior:** Africanized bee colonies swarm 3–4 times/year, often with one or more afterswarms following the primary swarm. European colonies in the tropics had no history of swarming. Swarm cells appear quickly in Africanized bees as colonies expand.
- **Absconding:** A common occurrence in Africanized bees (up to 30% of colonies) once the nectar flow stops; absconding does not normally occur in European bees. Continued disturbance (such as too frequent inspections) elicits absconding in Africanized bees.
- **Supersedure:** Rates of supersedure are higher in the tropics (>40%). We do not know the extent of differences that occur between the two races.
- **Requeening:** Difficulty of requeening and a lower success rate can be an indicator of Africanized bees. Normally 10% of queens are lost in requeening but a 40% or higher failure rate is reported when European queens are introduced to Africanized colonies. Africanized queens are difficult to find but respond rapidly to smoke by leaving the brood area (often going to the inner cover).
- **Time of mating:** Most Africanized bees mate after 5 P.M. while European queens mate before this hour.

Defensiveness Differences

There are major differences between European and Africanized bees in colony defensive behaviors. Africanized bees bounce off the face of the bee veil, respond to disturbances faster, defend with a greater proportion of bees, stay alerted longer (even the next day and sometimes for several days) and will pursue outside the colony a longer distance. (Figure 4–3)

- **Stinging incidents:** Accidents involving animals, persons in nearby houses, or people passing through or by the apiary occur frequently with Africanized bees. Few non-provoked instances occur with European bees and any stinging instances are likely to be confined to within the apiary itself.
- **Entrance guarding behavior:** Africanized bees attack other bees at the entrance and fighting is more commonly seen at the colony entrance. Africanized bees tend to use more propolis at their entrance. Africanized bee guards keep the wings elevated off the abdomen while European bees fold their wings on top of the abdomen.
- **Response to smoke:** More Africanized bees leave the hive and fly about when a colony is smoked. Inside the colony more bees can be seen filling honey stomachs from honey stores. Eventually the smoke seems to have no effect. European bees fan vigorously to expel smoke from the hive whereas Africanized bees exit the hive when it is smoked.

Colony Differences

Sometimes Africanized bee colonies are highly defensive and at other times they are not. Here are some more relative differences.

- **Foraging behavior:** Analysis reveals a number of differences including a more direct flight into the hive entrance without landing on the bottom board, a more rapid flight appearance and more pollen collection by Africanized bees.
- **Running behavior:** This unusual behavior is especially common in Africanized bees. When the colony is opened, Africanized bees run off brood frames and hang as a festoon of bees at the lower margin of the frame. (Figure 4–4) In the hive, they run around (parade) the hive box and inner cover.
- **Cell cappings:** In Africanized bees, cell cappings are thinner. It is more difficult to uncap such frames compared to frames capped by European bees.
- **Length of developmental stages:** Africanized bees have a more rapid development. Egg to adult takes 18.5 days for workers and 15 days for queens versus 21 and 16 days in European bees. Drone development is the same in both bees.
- **Brood production:** European bees are slower to respond to a change in forage conditions and build brood and adult

Figure 4–4. A field characteristic. Africanized bees running off comb.

populations slower. Africanized bees seemingly "explode" overnight and have fully two times the amount of brood once resource conditions improve.

- **Length of life:** There is a considerable difference between the two bee races. Africanized workers live 12–18 days during favorable foraging conditions, 15–38 days under dearth conditions. Both are considerably shorter life spans than for European bees.

With these number of differences, the Africanized bee colony seems different in the field. It looks and behaves differently outside the colony entrance; other differences are readily evident when the colony is opened and inspected. It is a different bee from the one we know in the U.S. and it is a distinct bee from the one beekeepers of various countries in the Americas are used to.

The greatest behavioral difference, and the one we hear most about, is colony defense. We often notice a delay in manifestation of defensive behavior when the Africanized bee population first moves into an area. Al Dietz, Africanized bee specialist and retired Professor of the University of Georgia, found "incidences of intense defensive behavior have not occurred until about 1½ years after the initial discovery" in Mexico. He called this delay in defensive behavior "a most remarkable phenomenon."

In Panama, Belize and other areas I have visited, I found a great deal of optimism among beekeepers during initial colonization. The reputation as a highly aggressive bee proceeds invasion and when bees do not immediately become defensive following colonization, the attitude becomes one of "well this bee isn't so bad—I can handle it."

The Africanized bee has proven not to be easy to identify. It has proven no easier to manage.

✧✧✧✧✧✧✧✧✧

So what's in a name? I prefer to use the term "Africanized bee" only because it has come to mean the African bee in the Americas, the neotropical African bee. Use of the term "tropically-evolved" bee, in contrast to "temperate-evolved," helps designate the bee race *scutellata* that developed in Africa. Such terms acknowledge that the bee may now be in the process of being modified in the American tropics. I and others often interchange use of the names "European," "European-race"or "temperate-zone" bees to designate the several bee races originating north of the Mediterranean. These bees have been transported most extensively throughout Europe, and into the Americas by beekeepers. Such bees would be the origin of what some call the "American bee" that we are accustomed to in the U.S.

Whatever we call this bee, it has certainly fostered many different reactions. The initial response of beekeepers to arrival of this bee into the U.S. has been the same as those expressed by beekeepers in Panama, Costa Rica and our neighbors in Mexico. We have much to learn from the experiences of fellow American beekeepers who have had to learn to deal with Africanized bees and identify its many variations.

5

Myths and Barriers

A tremendous number of stories about Africanized honey bees abound. When I give oral presentations about Africanized bees, I like to discuss an arbitrary sequential myth history of the Africanized bee. Myths illustrate that our perception of the Africanized bee, and the magnitude of the problems faced by beekeepers, has changed since we first heard about the killer bee.

For example, THE FIRST MYTH was:

"They are too far away to bother about."

This was truly the case in the 1960s and early 1970s when TIME magazine, the NAS team and *National Geographic* first informed us about "those fiery Brazilian bees." The media really loved killer bees.

With little scientific literature to counter sensationalized stories,
THE SECOND MYTH became:

"There is no real problem, only media hype."

Hollywood had fun with movies such as *The Swarm*, *The Savage Bees*, *Terror out of the Sky* and *The Killer Bees*. How many of these classics have you seen?

The USDA and University research programs started investigating details about Africanized bees. We learned about absconding, a relatively uncommon behavior of European bees, an astonishingly high swarming rate, seasonal population cycles that were not spring-summer-fall-winter and Africanization of European bee colonies. Dr. Taylor forced us to consider colonization potential with his 1977 *Bee World* paper (Chapter 1, pages 5–6) predicting eventual spread throughout the Americas "in our lifetime." (Chapter 2, page 24)

By the late '70s the THIRD MYTH came to be:

"If they really are a problem we can stop them at the isthmus."

Perhaps this myth developed from our John Wayne "We'll cut 'em off at the pass" mentality, or the viewing of too many western movies—

they were certainly better than the several killer bee movies Hollywood gave us! This is also when I came into the story with my initial visit to Panama at the same time Africanized bees arrived. (Chapter 6)

Alas, the tiny honey bee moved too fast for even the mighty U.S. military. The impenetrable Darien jungle of the isthmus of Panama did not slow Africanized bee spread.

The scene kept shifting northward as the Africanized bee continuously advanced through the Americas, getting dangerously closer to the U.S.

THE FOURTH MYTH was really a multiple choice option:

"The bee will never make it through Mexico" because:

a) no visitor ever gets through Mexico healthy;

b) the bee will be diluted out by the extensive European bee population; or

c) the bee barrier plan (see later in this chapter) really will work.

Mexico did not prove to be the Waterloo for Africanized bees we hoped it might be. The division within the industry, the difficult terrain, and the problems getting a Mexican barrier plan implemented (as realized earlier in Panama) only served to divide our forces while the Africanized bee population continued advancing northward.

We were learning a lot about Africanized bees from the cooperative U.S./Mexico research/barrier program. As we observed bee advancement northward through Mexico, it was obvious the Africanized bee was not "playing by our rules."

As the fourth myth dissolved we came to a FIFTH MYTH:

"We will stop them at the border."

Alas we should have consulted the U.S. Border Patrol and Drug Enforcement Agency (DEA) on their success rate in stopping the influx of illegal aliens and drugs. Just as they have not been able to halt human or drug invasions, we were not able to halt the Africanized bee invasion at the border.

Now we could study Africanized bees first-hand in Texas.

Another myth by the wayside! But we did not despair, the AHB (now we could refer to it by only using the acronym) was surely subject to the SIXTH MYTH:

"We can deal with AHB in the U.S. and eliminate undesirable characteristics by interbreeding them with European bees."

Do I need to point out that this, too, is merely a myth? Who is going to do all this interbreeding? Who is going to pay for it? How many beekeepers will regularly requeen their managed colonies? Why will interbreeding work in the U.S. where it has not in any other location where Africanized bees occur?

Such issues are troublesome but easier to answer in the abstract. Now, finally would we have to deal with the reality?

Well no, we still promulgated a SEVENTH MYTH:

"The Africanized bee will encounter a very large, locally-adapted feral population, along with very large numbers of managed colonies of gentle European-race bees in the U.S. Bee colonies can be requeened if they become too aggressive."

Do you still believe in this seventh myth? We may have had a large feral population but is this still true after Varroa mites? Frank Eischen in Texas, Rob Page in California and Jerry Loper in Arizona have projects in which they systematically survey for feral bees. All have found dramatically decreased numbers of feral colonies with Varroa mite pressure.

Can we reverse the gene flow here in the U.S., though it has not been done elsewhere? Research from Mexico clearly documents the one-way transfer toward African genomes. Will the bee finally reach its geographical limit, permitting U.S. beekeepers to maintain a distinct population of managed bees, with desirable characteristics north of a transition zone? Will U.S. beekeepers continue beekeeping as we now know it?

We will see!

What have we actually accomplished? We have:

- very competently tracked the spread of Africanized bees,
- learned how to ID bees better,
- learned a lot about bee biology,
- developed and discarded a lot of myths.

As the Africanized honey bee has spread throughout the Americas, we have repeatedly witnessed gene flow favoring African bee genetic material.

Everywhere there has been

Africanization of managed bees. Repeatedly, once the feral Africanized bee population arrives in an area, mating of European queens by Africanized drones, and to much lesser degree, usurpation of colonies by Africanized swarms, has resulted in gene flow toward Africanized bees away from European. Africanized drones take over by sheer numbers, not by any special mating advantage. Changeover is rapid, even in areas such as Venezuela and Mexico, where a large, modern beekeeping industry existed prior to invasion of African bees.

Africanization occurs but there is virtually **no Europeanization of feral bees.** Nuclear and mtDNA analysis demonstrate the small amount of Europeanization of the feral Africanized population that occurs as Africanized bees colonize areas with managed apiaries of European bees. The existing European-race population has little impact and European genetic material disappears rapidly.

Will history repeat itself in the U.S.?

OTHER AFRICANIZED BEE MYTHS THAT AREN'T TRUE

- African bees "get up earlier and work later" than European bees.

NOT TRUE

- Africanized bees produce "killer" honey, much stronger and more dangerous for humans.

NOT TRUE

- Africanized bees changed from *adansonii* to *scutellata.*

NOT TRUE

(Although the name changed, there was no change of bees.)

- Africanized bees are "sexier" and have a mating advantage over lazier, less aggressive European drones.

NOT TRUE

(The Africanized honey bee mating advantage appears to be a numbers game, not some special sex attraction of the Africanized drone to European queens.)

- Africanized bees were brought to the U.S. before 1990.

TRUE

(But the feral population, which is our major reason for concern, did not arrive until 1990.)

- The Africanized bee has a "killer instinct" and will attack without warning.

NOT TRUE

(Stinging incidences may be sudden and focus on innocent persons or animals but stinging behavior is defensive; bees are responding to some stimulus.)

- The sting of the Africanized bee is "more potent and more dangerous" than that of European bees—or, in another version, Africanized bee venom is very potent and a single sting can cause death.

NOT TRUE

(Although minor differences occur, bee venom is essentially the same whether injected by a European or Africanized bee. A single sting from one bee, wasp or ant can cause death in an allergic person).

- The Africanized bee has no known enemies.

NOT TRUE

(The gorilla, *Gargantua terrilibus,* [Figure 5–1] is not the "magic bullet" enemy of Africanized bees as "suggested" by Mike Burgett and John Ambrose in a 1976 *American Bee Journal* article.)

Figure 5–1. Biological control "gorilla"—a Burgett-Ambrose myth. (From: 1976 American Bee Journal 116:22)

- The Africanized bee is bigger, darker, uglier.

NOT TRUE

(The Africanized bee is actually 10% smaller. Populations exhibit wide variation in body color, though drones tend to be darker. Hive and field behavior and other morphological features and biologies vary within the population of Africanized bees.)

Do you have other myths to add to this list? The Texas A&M Africanized bee website http://agnews.tamu.edu/bees/quaran.htm features a myth quiz. I predict that by the time you finish reading this book, you will be able to compile a much larger listing.

Figure 5–2. The real thing—defensivenes is no myth!

Bee Barriers

It is human nature to seek to find a solution when confronting a problem or challenge. Pessimistic beekeepers (the glass is half-empty) hearing about the Africanized bee, might think that little can be done. Optimists (the glass is half-full) think that something can be done to stop Africanized bee population advances.

One widely-promoted approach (or myth, depending on your point of view) was the proposed establishment of a barrier to stop the spread of the Africanized bee into the U.S. The 1972 NAS report, was probably the genesis of many of the barrier proposals. In their summary, the NAS team report stated

> "To control the northward spread of this bee will probably require action on the Central American isthmus, perhaps within the next four to six years—action the nature of which will depend on the results of experimental work not yet begun. There is no known geographic or climatic barrier that will prevent the spread of the Brazilian bee into North America."

The NAS committee had three suggestions on how to make an effective barrier. They believed that swarms of Africanized bees could "probably" be captured if bait hives with a suitable attractiveness were distributed in a strip across Central America and poison bait stations could be employed to halt the movement of Africanized bees northward. They quickly acknowledged that the methodology to put such options into use was not yet available.

Barrier in Panama

Panama was probably the first logical location where a barrier could be created to stop the advance of the Africanized bee. In Panama, the U.S. had a large military presence of over 20,000 troops, to protect the Panama Canal. In addition, the Darien of Panama is acknowledged as one of the wildest, most impenetrable jungles in the world; no road connects North and Central America to South America. Only a few native Kuña and Choco Indians live in the Darien jungle, and they only live along the coast and on a few of the rivers inland.

As he witnessed spread of Africanized bee from São Paulo to Rio de Janiero, Nogueira-Neto, the Brazilian bee scientist, wrote "we can only hope" the Africanized bee will "encounter effective enemies" in the Amazon or "in the Darien jungle of Panama." While witnessing Africanized bee spread through the Amazon, Mark Winston reported that "many Central American beekeepers believe the rain forest of eastern Panama might stop the bee's advance." Unfortunately neither the Amazon nor the Darien jungle stopped Africanized bee advance!

I do not know what barrier plan myth you have heard or if you have a favorite. One of my favorites was that a giant net would be suspended at the Panama Canal (by the U.S. military of course) to intercept all swarms. Two other equally wild solutions in print were: create a high wall of gas-fueled flames along the Panama Canal or create a wide radiation belt with nuclear weapons through the jungle. Surely technology would be the answer!

Another barrier suggestion was a variation on the "spray-and-pray" method of killing pest insects practiced by some farmers. The idea was to spray a 80 km (50 mile) swath of the Darien jungle at intervals to kill all migrating swarms of honey bees. I do not know the origin of this plan but Professor Kempff Mercado in Bolivia, among others, mentioned the idea of the "insecticide curtain" plan. I suppose the concept of an insecticide barrier was considered because pesticides were then creating great harm to beekeepers in the '70s.

Thankfully, neither nuclear or pesticide barrier plans were implemented in Panama though, incredibly, the "scorched earth" idea surfaced again as the bees approached Guatemala. Espina in his Spanish language book, *Beekeeping of the Assassin Bees*, wrote about using highly toxic compounds applied to cotton, to produce a barrier by "...intensively treating with insecticides...to impede the access of the Africanized bees to Mexico."

I do not know if the cotton industry failure in Guatemala (due, of course, to the high cost of production from extensive use of costly insecticides) preceded bee arrival there or clearer heads prevailed but, thankfully, a pesticide barrier was never implemented—in Panama or elsewhere.

Beekeepers then suffering repeated pesticide kills in the U.S., Guatemala, Brazil, and elsewhere might have wondered if their bees were not inadvertently being used in such experiments.

Genetic Barrier

Another barrier suggestion, originally expressed in the NAS report, was a genetic barrier. The committee recommended immediate feasibility-testing to demonstrate whether a genetic barrier would work in combination with a program to select a "non-aggressive, non-swarming, and non-migratory bee." They recommended selection should be developed "in several lines of various origins, such as from bees from northern and southern Brazil and Africa." Of course, Panama was where the barrier itself would occur.

The late Dr. Marshall Levin was head of the USDA research program in apiculture at the time of the NAS report. Although he was not in total accord with the NAS report, he did commit to studies on some of the research recommendations. As for the barrier, he is quoted as saying: "Let us do what we can to provide a natural genetic filter, instead of a genetic barrier, by encouraging and developing beekeeping at an expert level in the countries in between" (meaning between Brazil and the U.S.). He continued "I think this is probably the most positive action we could take, and some of the countries—Panama and Costa Rica in particular—are already doing some of this on their own."

Although the Brazilian bee scientists found much to disagree with in the NAS committee report (and produced a 35-page rebuttal), Professor Kerr agreed with the genetic barrier recommendation of the committee. He wrote "I think the idea of making a genetic barrier in the Panama Canal is a very sound idea, has a scientific basis, and is feasible. If some of the bees crossed the barrier, it would be easy to get rid of them; just alert the beekeepers to check the most aggressive bees and kill the queen."

Alas, time was too short and the bee too prolific for a barrier plan to be implemented in Panama. Well-funded and equipped Panama Control Commission "Bee Swat" teams located and removed 175 swarms in the 50 km^2 (20 mi^2) Panama Canal Zone area the first year alone. During the next 36 months, 1000 additional swarms/colonies were removed in the Zone. An additional 1000 were removed from zone military bases, and over 6400 from Panama City during the same time frame.

Despite the extensive search and destroy effort, the bee invasion resulted in several stinging incidents and did not halt or slow the spread of the bee. (Chapter 7) Was the problem a lack of a workable plan and failure to appropriate sufficient money to implement a barrier or was the idea of a barrier bad science? Panama as a barrier was a "bust"!

Mexico Barrier

With the invasion of Panama and continued spread of the Africanized bee population northward through the remaining Central American countries, interest in a bee barrier shifted to Mexico. It was closer to home; there was a sizable (estimated 2.5 million) managed bee colony population in Mexico; the climate is subtropical rather than tropical; and programs could be conducted more easily as Mexico had a sizable Ministry of Agriculture (SARH—Secretaria de Agricultura y Recursos Hidraulicos) with a history of helping beekeepers. Also, politically, spending money in Mexico could be more easily justified in the U S.

A Mexico-U.S. cooperative effort had another advantage—a successful joint project combating another insect, the screw worm fly, a serious pest of cattle and other livestock. The screw worm maggot invades wounds and birth placentas (navels) of range livestock and then proceeds to eat their flesh, slowing growth and even causing death. The fly has been eliminated from the U.S. and northern Mexico using the release of lab-reared, sterilized adult male flies (the sterile-male technique of insect control). The joint USDA-SARH screw worm project maintains a barrier at the isthmus of Tehuantepec in southern Mexico.

In 1985 and 1986, several meetings of Mexican/U.S. regulatory and bee scientists jointly devised a Bee Regulated Zone (BRZ) plan for the isthmus of Tehuantepec (Figure 5–3) covering an area 225 km long by 170 km wide (140 by 106 miles). The BRZ was not contemplated as a barrier to

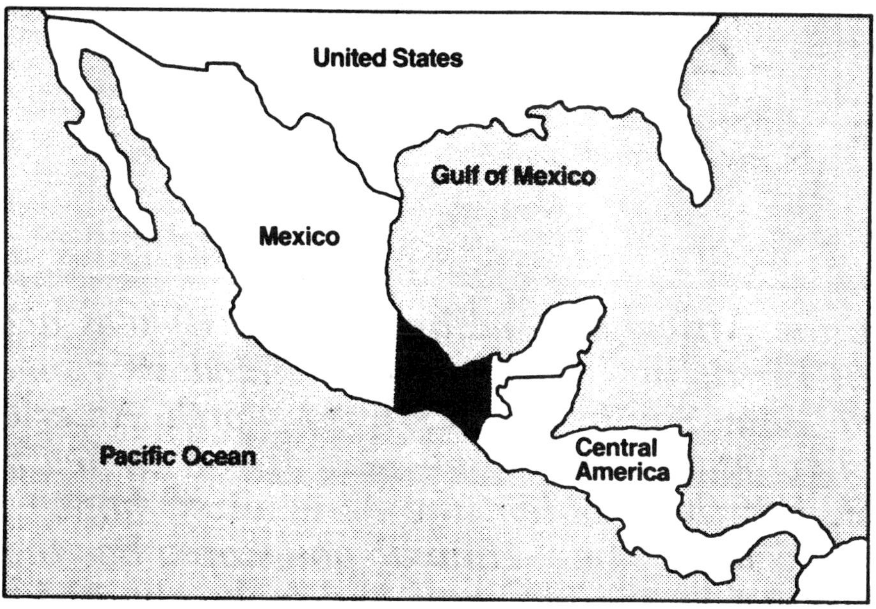

Figure 5–3. Original Bee Regulated Zone (BRZ) at the isthmus of Tehuantepec. (From: American Bee Journal 126:729)

bees per se, but rather as a zone where the undesirable characteristics of Africanized bees would not be allowed to cross—a biological barrier.

The BRZ plan was designed to take advantage of the extensive surveillance program already in place to prevent movement of the screw worm fly northward. The plan included a quarantine area prohibiting movement of bee colonies out of or through the isthmus. Swarms would be captured in 141,000 pheromone-baited, cardboard traps. Swarms that did not settle in a trap would be found and eliminated. Detection of such swarms depended on local residents notifying authorities of swarm locations; their incentive was a $1.00 bounty for each swarm they reported.

Another component of the plan was to teach the beekeepers of the area how to better keep colonies, including how to requeen bee colonies. Most beekeepers in the region did not have much prior training in the care of colonies. Some of the colonies were rustic hives, without moveable frames. Queens would be marked and the beekeepers would be encouraged to check their colonies frequently to insure that a swarm did not invade and Africanize the colony. Colonies would be fitted with queen excluders to help prevent loss of swarms; in all, some 39,000 colonies were to be included as part of the BRZ. Demonstration apiaries south of the BRZ would be established so beekeepers could observe Africanized bees and learn how to recognize if Africanization was occurring.

Within the zone individual flower-foraging workers would be captured and identified to be certain colonies remained European. Also special attention would be given to drones. Nearly 16,000 drone traps were to be placed near apiaries to attract and kill Africanized drones. One thought was to design a trap that would contaminate Africanized drones with an insecticide and then release them to carry the poison back to the parent Africanized colony. Beekeepers would be encouraged to produce more European drones in their managed colonies.

The BRZ program was labor-intensive. It was to include over 1100 employees, 220 vehicles and projected to cost about $8 million, shared between Mexico and the U.S. By comparison the screw worm fly program annual budget was $32 million.

Rapid movement of the bee into Mexico and a lack of political support for a bee barrier within the U.S. delayed implementation of the BRZ. I remember our national bee meetings at the time as particularly acrimonious with wide differences of opinion expressed. Many of the scientists, who were competing for what was then diminishing research funding to study the Africanized bee, did not believe the large expenditure of money, with mostly unproven technology, such as pheromone or drone trapping as well as the concept of creating a zone of high European population colonies and drones, stood much chance of meeting barrier expectations.

For example, Roger Morse was quoted as concluding that "talk of genetic dilution is ridiculous...it has not happened in the 31 years since the bees arrived in Brazil and there is no reason to think it will happen in

Mexico." Orley Taylor felt "We cannot Europeanize these bees. They are going to reach us essentially unchanged." TIME magazine had a cartoon of a brick wall bee barrier (a la Berlin Wall).

Revised Barrier Plan

A revised $6.3 million, 150-technician, 38,000-colony, and 73,000-pheromone-trap program, termed the "Africanized Bee Cooperative Program," was eventually implemented in September, 1987. Unfortunately the original barrier at the Isthmus of Tehuantepec was already invaded by Africanized bees so a new zone farther northward where the Sierra Madre Mountains channeled the invading Africanized bees along the Pacific or onto the Gulf of Mexico coasts was designated. The Pacific area was 33 km by 117 km (21 miles by 73 miles long). The Gulf Coast area was 200 km by 75 km (124 miles wide by 47 miles long).

The revised program was no longer labeled a bee barrier. It was intended to slow down the northward movement of the invading bees by "incorporating various technologies to reduce the advancing Africanized honey bee population, while increasing that of the European honey bee." TIME magazine was more favorable to this plan than they had been to the earlier one; many scientists remained skeptical.

The basic technique proposed for the Cooperative Program was to establish trap lines of bait hives and bait stations where there was little beekeeping in order to monitor Africanized bee spread. Plans called for increasing the number of bee hives and regularly requeening the managed colonies within the zone, along with transfer of rustic hives to standard equipment to create a buffer of managed colonies. A bounty for captured swarms was also included.

Unfortunately, discovery of Varroa mites in the U.S. held up shipment of U.S.-reared queens into Mexico thwarting part of the plan. There were problems verifying the authenticity of swarms for the $1.00 bounty and this plan lacked adequate provisions for compensating beekeepers for their contribution. Serious division among Mexican beekeepers and their representative associations also occurred which eroded program support and beekeeper cooperation.

Program objectives eventually were restated as an attempt to monitor and slow Africanized honey bee movement northward. This became increasingly impossible as the Africanized bee quickly passed through the Pacific zone. When the program was terminated in 1990, objectives were more generally listed as an improved understanding of techniques to monitor Africanized bee movement and gaining a better understanding of what might happen when the bees reach the U.S.

Some 80,000 swarms, 75% of which were identified as Africanized, were captured in the joint program. Al Dietz credits the delayed advance of the Africanized bee into the U.S. (at least relative to Taylor's prediction in 1977) to the joint USDA/SARH program in Mexico, though there is no evidence to support this opinion. Mark Winston's more modest assessment in 1992 was that the BRZ "did not stop or even slow the Africanized honey bee, although it did succeed in providing information that will be useful in controlling these bees when they colonize the U.S."

Failure of the Mexican barrier can be attributed to operational procedures, the great difficulty of coordinating a program between two different countries' governmental agencies, the logistics of working in remote, inaccessible areas, but, most directly, to the Africanized bee itself. Swarms and, in the final analysis, the concept that a genetic filter or barrier, whereby a large managed population of European bees would modify the genetic composition of Africanized bees, have not proven to be true. Africanized bees were able to build up large populations both inside and outside the control zones and move before monitoring/control programs could be implemented.

Canadian Barrier

With the appearance of bee mites in the U.S. in the mid 1980s, Canada elected to close its border in 1987 to package bees coming from the U.S. This action had "enormous" impact affecting over "half the 600,000 to 700,000 managed Canadian colonies" according to Mark Winston. By 1997, mites had become more widely distributed in bee colonies in Canada but the border remains closed. The 1990 arrival of Africanized bees in the U.S. has now become the major deterrent to lifting the ban of importing bees into Canada.

Barriers Revisited

Would a barrier have worked in Panama any better than the barrier in southern Mexico? Why were so many people thinking the U.S. military could do something at the Panama Canal? It would have been cheaper, as the isthmus there is only 80 km (50 miles) wide compared to 241 km (150 miles) width in Mexico.

The Panama Canal is truly one of the engineering wonders of the world. It is 70 km long (43 miles) and runs Northwest (Atlantic) to Southeast (Pacific). The area known as the Canal Zone, under U.S. jurisdiction since 1903, is an area extending 8 km (5 miles) from either side of the canal, except where the cities of Colon, at the North or Atlantic side, and Panama City, at the southern or Pacific Ocean side border the canal.

A five-mile-wide expanse of jungle alongside the Canal was left undisturbed to make land attack or sabotage of the waterway difficult. It is one of the most dense jungles in the world, except where it has been interrupted for activities connected with maintenance of the Canal, for training of U.S. military troops in jungle warfare, and by the encroachment of human populations, especially on the Pacific side.

Part of the barrier plan was to requeen colonies of bees showing Africanization characteristics. For there to be beekeeping, you need human settlement. Communities and settlements occur outside the canal

Figure 5–4. Panama Canal with jungle buffer from Barro Colorado Island.

five-mile limit along a parallel trans-isthmus roadway to the south of the Canal along with some authorized (and unauthorized) settlement along the shores of Lago Alejuela. Technically the area around Lago Alejuela is Indian territory and most of the immediate shore frontage is part of Chagres National Park.

On the Atlantic side of the canal, small communities occur along the coast but the majority of this territory is an autonomous native Indian (Kuña) area. The Pacific side, in contrast, is densely populated. Panama City, at the Pacific terminus of the canal, is home to about half of the 2.5 million inhabitants of Panama.

No beekeeping existed in the Canal Zone. Two or three years prior to Africanized bee arrival, one commercial beekeeper, General Paredes, established about 400 colonies in a series of apiaries along the highway south of the canal. A few hobbyist beekeepers, such as Sam McClymont (Chapter 6), had apiaries in some of the suburbs of Panama City but the number of actively-managed hobbyist colonies probably numbered less than 100. Since swarms were not a feature of honey bees in the tropics, there were virtually no feral colonies of honey bees anywhere in Panama.

Clearly, the conditions suggested by the NAS committee for a barrier such as killing wild colonies and requeening aggressive colonies inapiaries would not have been met in the Canal Zone. The dense jungle made locating feral colonies/swarms or area spraying entirely ineffective. There were too few managed colonies to make any requeening program worthwhile; lack of adequate numbers of managed colonies is one of the factors cited for the Mexican plan failure.

Orley Taylor pointed out in 1988 that fewer than 500,000 managed bee colonies (and virtually no feral colonies), would be encountered by the Africanized bee population as it colonized countries from Southern Brazil to Texas between 1957 and 1990. The majority of Mexico's estimated 2.6 million colonies were located in the Yucatan and highlands, which the bee bypassed on its route to southern Texas.

Beekeepers I talked to in Panama and Costa Rica wanted to move their bee colonies to locations (highlands and/or islands) where they might suffer less impact from Africanized bee arrival. Many beekeepers did not do so because their first impressions of the bee were false ones, leading to complacency and less incentive to follow through. Mexican beekeepers likewise, did not wish to leave colonies in the bee-regulated zone as a barrier, preferring to move their colonies to the highlands and delay Africanization.

Will it be any different in the U.S.?

Robert Kent, based on his beekeeper survey in Peru, was not in favor of a barrier plan. He found the Africanized bee not to be a major problem in the view of the majority of the beekeepers in Peru. He suggested we "should dedicate these funds to extension education for beekeepers to help them and the public prepare for the inevitable arrival of the hybrid" (i.e. Africanized bees), which he found important to success among Peruvian beekeepers.

Ultimately, baiting techniques were not effective when the bee arrived in Panama and they still were not adequate in Mexico. Attractive baiting remains an elusive hope (myth) as a control measure for any insect, including honey bees.

And so we are left with myths—but do not include barriers, physical or genetic, in our prognostications.

✧✧✧✧✧✧✧✧✧

A barrier to Africanized bee northward movement at Panama or Mexico was not successful, and even today, with our present state of knowledge, would not be feasible. A barrier, like much else we have proposed as we sought to tame Africanized bees, was a myth. Now we have to deal with the actuality—Africanized bees within the U.S. We need to learn from the experiences of others and to continue to build on our knowledge base concerning Africanized bees. Knowledge will be the best means of countering barriers—and our myths—that are ill-founded with little basis for success. Myths have not been, nor will they be, enough to combat the invader!

6

First Impressions and First Encounters

My initial exposure to Africanized bees and tropical beekeeping was in the country of Panama. Beekeeping in the tropics is not so strongly governed by seasons as in temperate areas. Now, following arrival of Africanized bees, both the bees and beekeeping are different from the U.S.

Panama, 9° north of the equator, connects the continent of South America to Central and North America. It extends 680 km (422 miles) from Colombia on the east to Costa Rica and the rest of Central America on the west. (Figure 6–1) It is nearly east–west in orientation, unlike the remaining Central American countries, which are north–south geographically.

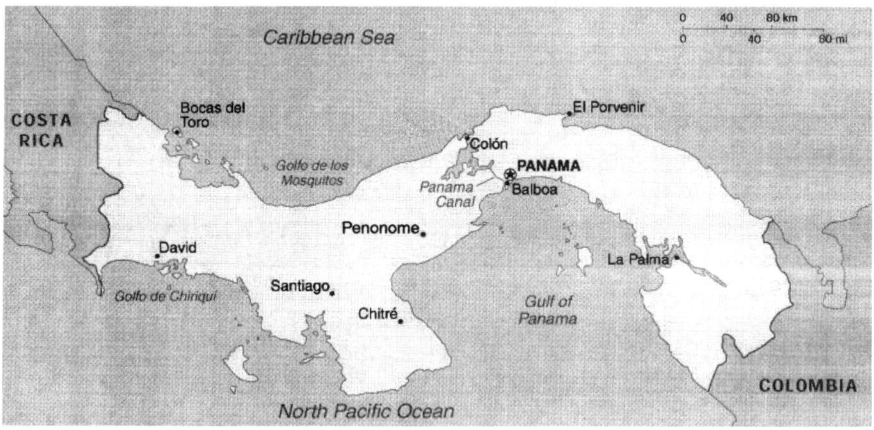

Figure 6–1. Panama. (From: The World Factbook)

Panama, about the size of South Carolina (77,080 km^2 – 29,752 mi^2), is a narrow isthmus, a mere 50 to 120 km (31 to 75 miles) wide. It is mountainous with a central mountain range that reaches an elevation of 3374 m (11,070 ft) dividing the country into Pacific and Caribbean portions. The population is about 2.5 million.

Panama is less rural than many of the countries of Central America. An estimated 40% of the total population was listed in agricultural occupations which accounts for 20% of the GNP of the country. Agriculture uses 30% of the land—6.6% for crops, with 15% devoted to pasture for cattle and sheep. Exportable commercial crops include tropical ones such as bananas, sugar cane, oil from oil palm, rice and coffee. Other crops, produced mostly for in-country use, are beans, cocoa, corn, vegetables, citrus and other tropical fruits, milk, eggs and poultry. Rice and corn are staples, but the diet of the typical citizen is quite varied.

Honey was a minor product. There was export of some honey to Europe (Germany and Switzerland) and an expanding honey market within the country before Africanization. The local market is protected with a high tariff on imported honey, though contraband honey from Guatemala and Costa Rica is a serious problem, especially in the western province of Chiriquí where the large honey production of this region was most directly in competition with the illegal imports.

Panama Discovers Africanized Bees

Africanized honey bees officially reached Panama in February 1982 when regional apiary inspector Alberto Silva observed bee swarms in Sante Fe in the Darien region. Swarms were not common in his isolated area and when he hived this swarm, he noticed that the foraging bees flew directly into the colony, a behavior different from his European bees.

A Ministry of Agriculture (MIDA) veterinarian, Dr Ramón Boyd, collected an Africanized bee swarm on the outskirts of Panama City in March of the same year; in the Panama Canal Zone the first Africanized bee swarm was captured in the same month. Bee swarms had not been reported in the Canal area for at least 16 years preceding this swarm. By December of 1982, Africanized bee swarms were observed in the westernmost Pacific-side province of Chiriquí, a distance of 450 km (280 miles).

Movement of the Africanized bee population across Panama thus was extremely rapid. Spread from Sante Fe to Panama City, a distance of 250 km (155 miles), and on to David, another 450 km, documents total expansion of 700 km (435 miles) in 12 months. The following year, 1983, swarms began arriving on islands off the coast. Tobago, a popular day-excursion island 12 km (7.5 miles) off Panama City, had its first swarm in March of 1983. By 1983 there were numerous reports of swarms on ships transiting the Panama Canal.

Beekeepers in Panama had heard of Africanized bees (las abejas malas [the bad bees], or las abejas bravas [the fierce bees]) and they were ready for the invasion. At least a couple of the commercial beekeepers had visited beekeepers in Brazil. But the bees did not initially live up to expectations. The projected major impact expected for the next dry season following arrival did not occur. The rapidity of movement from one end of the country to the next, led some to the impression that "the bees were merely passing through."

Beekeeping in Panama

Although honey bees had been introduced into Panama in the early 1800s, Panama did not have a large beekeeping industry when the Africanized bee arrived. In 1971, there were only an estimated 3200 bee colonies in Panama. However in the five to seven years preceding the arrival of the Africanized bee, the number of European-race bee colonies in Panama had increased three-fold.

This change was due largely to a foreign aid program from Germany (Gesellschaft für Technische Zusammenarbeit or GTZ), a development program promoting beekeeping in Panama and several other Central American countries. Development funds were used to build a training facility and a beekeeping supply manufacturing/beeswax rendering plant at Divisa in the center of the country at an agricultural school. Professor Luasces, trained in Cuba, coordinated the educational program at this facility which was responsible for training all or most of the beekeepers in the country.

First Experience

The state of Delaware, through the Partners of the Americas and the University of Delaware, has a cooperative agreement (convenio) with Panama. Through these programs, I have been able to visit beekeepers in Panama and host Panamians in the U.S. on numerous occasions. The University has helped train faculty members of the College of Agriculture (Facultad de Ciencias Agropecuarias) including Javier Almillategui and Baltazar Gray, who have cooperated in joint extension/research.

In 1982 the University of Delaware received funding to strengthen the faculty in international activities (termed Title XII). Panama was targeted for a cooperative research/educational program. Interamerican Development funds were then widely available and Panama was in the process of completing an entirely new facility to teach agricultural sciences in the western city of David, Chiriquí Province, the major agricultural region. The University of Delaware applied for funds to strengthen the educational background of the agricultural science fiaculty.

On my first trip to Panama I quickly understood that honey bee colonies were valuable livestock to the rural campesino.[1] Campesino beekeepers kept their bee colonies in their yards, alongside other domestic animals like chickens, pigs and cows. For the subsistence farmer, a chicken represents the checking account of the family. When the family needed something like seeds or a hoe, they traded or sold a chicken. A cow or pig represents the family savings account. When the family needed a lot of money for a funeral, wedding, or a piece of equipment, they could sell the cow or pig. Offspring from the pig or milk from the cow helped meet family expenses.

Bee colonies are like other livestock to the campesino family. Honey income helped meet regular expenses. If the family needed a lot of

---

[1]"Campesino," as used in this book, is a subsistence farmer—a small farmer in U.S. terminology.(Figure 10-3 page 135)

money they might sell a bee colony. There were always customers for a bee colony because swarming was not common in European bees in the tropics. The only way to start keeping bees was to buy an established colony or a nuc. A standard colony would usually get double the price we might expect to pay in the U.S.

Many of the bee colonies owned by campesinos were not kept in movable-comb hives, since most bee equipment had to be imported and therefore was expensive. The hives of rural beekeepers, called rustic hives, were often heavy, cumbersome boxes, though sometimes they were the same Langstroth and Dadant hive body size, only lacking movable frames. Whether movable comb or rustic, bee colonies were valuable and watched/guarded just like cows, pigs, or chickens. (Figure 6–2)

Figure 6–2. Bees in front yard before arrival of the Africanized bee. (From: American Bee Journal 122:112)

I met a number of part-time/hobbyist beekeepers in Panama in 1982. Many lived in the suburbs of Panama City, where 50% of the population of the country is concentrated. Typical of the beekeepers is the story of Sam and Greta McClymont. The McClymonts retired to Panama after living in south Florida where Sam worked as a charter boat captain. Before retirement both learned beekeeping in south Florida. They started with one colony, purchased from a neighbor in the Panama City suburb of Arraijan.

Some of the frustrations and challenges of keeping bees in Panama were evident in a 1969 article Sam wrote for *Gleanings in Bee Culture*. When I met the McClymonts in 1982 those earlier experiences were past history as their operation had grown to 350–400 colonies. Sam was called by customers "the sweetest man in Panama."

Even before I met the McClymonts, I knew I wanted to know them. I had seen their honey in a supermarket in Panama and the brand name Miel Mañana had me intrigued. I had quickly learned that the culture and philosophy of our neighbors to the south were different from our own. "Mañana" was a term that typified the more relaxed approach.

When you asked when someone was going to arrive or something was going to be done, "mañana" was often the response. "Mañana," meaning tomorrow was not to be translated literally, as "mañana" could be tomorrow or the next day or a week away. I wondered why would a honey be called mañana. Did that mean it was produced by really laid-back, relaxed bees?

Figure 6–3. The McClymont's label.

It turned out that Sam, an excellent, self-taught beekeper did have very gentle bees. He said he seldom wore his veil except when harvesting honey. We walked out from the cool, shaded patio behind his comfortable house in Arraijan, right up to the dozen colonies not 3 m (10 ft) from the patio and he reached his hand in front of a colony to show me his well-behaved bees. (Figure 6–4) He gently lifted the cover to demonstrate a partial super of honey, reminding him that he had to harvest that evening.

I observed Sam inspecting a colony and was impressed with how he moved slowly and smoothly with little wasted energy in his inspection. He had a smoothness to his movements while inspecting colonies that reminded me of the grace of a ballet dancer. He looked, satisfied himself that the colony was fine and then closed the colony without so much as a bump or jar. He used relatively little smoke and worked without his veil, though the colony was full-size in two boxes. He was truly at ease among his beloved bees.

I learned however, that Miel Mañana, their well-known brand, was not named for the gentle bees. When the McClymonts settled in Panama and were able to harvest honey the first season, Greta set about registering her product for sale. Here she ran head-on into the mañana philosophy. In frustration, she told Sam after one trying day dealing with government officials that if they ever got it licensed they would have to name it mañana. Well Greta finally did get it registered—and she appropriately called it Miel Mañana!

Who says beekeepers do not have a sense of humor?

Figure 6–4. Sam McClymont in front of his colony in 1982 (pre-Africanized bee invasion).

Beekeeping in the Interior

To the west of Panama City, in the interior, I found additional hobbyists around regional centers Divisa and Penonome, where beekeeping was taught at agricultural schools. However, the largest and greatest number of beekeepers were located in the western-most Province of Chiriquí, where most of Panama's agriculture is found. Four of ten commercial beekeepers were located here with an estimated 50% of the approximately 20,000 bee colonies in Panama.

On my first visit in 1982 I met five of the largest beekeepers. I would later meet the remaining five. Within five years following Africanization 8 of those 10 would all be doing something else and bees would become only a small part of how they earned money. There were no full-time queen breeders nor supply dealers in Panama in 1982, but several beekeepers raised queens for their own use. A couple of beekeepers sold a few of these queens, along with bee boxes and frames, mostly as nuc colonies. Foundation was available from the training facility in the center of the country (Divisa) where many beekeepers took their wax for rendering. Only necessary bee equipment that could not be made locally was imported.

Each of the commercial beekeeping operations demonstrated their individuality. Many beekeepers had taken advantage of a development bank loan program recently offered by the government of Panama to stimulate apicultural development. Funds were used to purchase hives and upgrade extraction equipment. I saw brand-new 30- and 50-frame extractors, imported from the U.S., still in shipping crates. Everyone had new hive bodies and frames stacked up awaiting division of colonies during 1982.

Let me introduce you to two of Panama's larger beekeepers.

Juan Malivern

Juan Malivern was the largest beekeeper in Panama. He was educated in Panama as an economist and worked for the vegetable growers' cooperative for many years before becoming chief agricultural loan officer for Chase Manhattan Bank. He became interested in honey bees, bought a few colonies 25 years ago and then decided he wanted to work outdoors, so he quit the bank in 1975 to go into commercial beekeeping.

Juan maintained a well-organized, economically-sound bee business in Chiriquí Province, one of four commercial beekeepers in the Province. He kept thorough apiary and individual colony records and regularly requeened those colonies that did not meet his expectations. He was unique among beekeepers in not utilizing the Development Bank loan program offered to beekeepers in the late 1970s because he wanted to first see what the Africanized bee would be like. He was confident he could handle a more defensive bee but during family visits in Brazil he had first-hand experience with Africanized bees and surmised (correctly) that loan repayment would be impossible with Africanization.

Figure 6–5. Juan Malivern inspecting a colony in 1987 (post-Africanized bee invasion).

I was very impressed with his colonies, his well-built and efficient honey house, and his management skills. He employed a full-time assistant and used hard-working day laborers which he paid well. His hives were all in top condition, painted with reflective aluminum, on similar, sturdy hive stands, which were well-supered when I visited him in the nectar flow season. It was a very professional-looking operation. None of this matters to beekeeping success of course—it merely is a reflection of the type of person he is. What did matter were his strong colonies, his knowledge of honey plants and his constant attention to management details. He was not adverse to trying something new which he carefully evaluated, keeping records to verify effectiveness of his trials.

In 1982, Juan was intently interested in queen rearing. He had the basics down and I spent some time going over details with him. He could quickly grasp the things I demonstrated. His wife Estrellita, an active partner in the bee business, was talented at grafting young larvae to queen cups. He intended to use apiary sites at higher elevations of 1000–1400 m (3300–4600 ft) for queen rearing; he was then at the point where he could requeen his 600–700 colonies annually.

Hermann Gnaegi

Another commercial beekeeper I met during my first visit to Panama was the President of the National Beekeepers Association, Hermann Gnaegi. Señor Gnaegi was of German descent and, besides his own production, was buying the honey production of a number of beekeepers and exporting it (along with fruit products and cashew nuts) to Europe where he had relatives in the import business.

His bees were located in Nata, in the center of the country. At the time he had 400–500 colonies. The majority of the colonies were on cattle ranch property (mostly his own), common in that part of Panama. His apiaries were neat and well-organized as was his honey house and entire operation. He had recently been experimenting with honey wine and we enjoyed his most recent fermentation while seated on his shaded patio. Alongside the house was an apiary; the honey house was in the backyard.

Señor Gnaegi wrote to Dario Espina in 1982 expressing the same optimism he voiced with me regarding the Africanized bee. At that time, Espina was writing a book about beekeeping with Africanized bees, (printed in 1985 as *La Abeja Africanizada* and the next year with a slightly different English title *Beekeeping of the Assassin Bees).* Gnaegi wrote:

> "Panama produces about 500 tons of honey a year. I visited Brazil at the beginning of this year and was convinced that the Africanized bees are not as bad as they are made out to be and I am not afraid of them. I hope they very soon crossbreed with ours in Panama because they are outstandingly more productive."

Señor Gnaegi had plenty of reason to be a gracious and optimistic host in 1982. He was the person largely responsible for beginning the export of honey to Europe (Table 6.1). With a significant BID (Development Bank) loan Hermann had recently purchased a new radial extractor and many new hive bodies. All the hives were standard Langstroth equipment.

Table 6.1

Year	Destination	Amount (kg)
1980	Switzerland	56,966
1981	Switzerland	65,866
1982	Switzerland & Germany	150,000
1983	Switzerland & Columbia	164,444
1984	Switzerland	17,777

Penonome School Visit

During this initial trip I visited an agricultural high school in Penonome, 120 km (75 miles) west of Panama City. The school proudly demonstrated their 12-colony apiary maintained just to the side of one of the classrooms. The school emphasized practical aspects of agriculture. Students came from all areas of the country but mostly were from the central provinces around Penonome. Student training included two years of instruction in raising of the animals and growing the crops, which the students did since there was little staff. The school, like many in the third world, needed to raise enough animals and produce to both feed the students and meet necessary school expenses selling surpluses.

The bee instructor had learned his beekeeping through the course at the regional agricultural school in nearby Divisa. He was assisted by the regional Ministry of Agriculture bee specialist for that Province. All students learned by participation. Honey was sold in the local market on market day, along with other produce and animals raised by the school.

All the bee equipment was relatively new, coming from the loan program of the development bank. There was a new extractor, a room to store equipment, about a half-dozen bee veils and other accessories. The colonies were all gentle, in excellent condition. They produced an average of 45 kg of honey (100 lb) in 1982 and 40 kg (88 lb) in 1983.

Africanized Bees in Penonome

My second visit to Panama was in 1984. Africanization had occurred for the majority of the bees in the Panama City area, in Penonome and in the rest of the central portion of the country. Africanization was not yet complete in the western Province of Chiriquí, being only the second dry season (when bee colonies store surplus) since arrival of the bees in this region.

On a return visit to the agricultural school in Penonome, I found only three colonies to open and inspect. The colonies had become defensive so they had been moved to the hillside behind the school. We donned protective equipment in the shade of the classroom in preparation for our inspection. Then we walked back into the brush about 200 m (650 ft) to the three colonies. In the company of the bee instructor and MIDA regional bee specialist, we approached the apiary cautiously. The instructor had changed to the large smoker typical of Brazil so it took awhile to fuel and fire it up.

The first Africanized bee colony I examined was a unique and unforgettable experience. The colony was not very strong, consisting of only one hive body sitting on bricks. (It is the colony on the cover of this book.) It was heavily smoked and when opened I was surprised at how the bees quickly exited and immediately flew around our heads. Removal of the first frame led to even more bees in the air. This was certainly a different reaction than I had experienced in 20 years of keeping bees.

Even more remarkable was the behavior of the adult bees. They became very active, running from the combs when we looked in the colony. The adults abandoned brood and, after a frame was out of the colony for a minute or so, it would be almost emptied of adult bees, except for a cluster of bees hanging off the lower edge.

My reaction was one of wanting to stay still. Bees exiting their hive "popped" at the veil and after a short time we each had perhaps 200 bees flying about our veils. I could smell alarm odor in the air. Bees were flying everywhere but they were obviously concentrating on us. As I looked into the colony, bees could be seen "parading" on the side wall and bottom board. Smoke did not seem to have much effect. The adult bees flying around us did not land and go back into the hive—instead more continued to depart from the hive.

I felt almost paralyzed. I did not want to move for fear I would be stung. It was almost a helpless feeling, one where I felt we had no control over the colony. It seemed to be an extreme reaction, brought on only by opening the hive and inspecting the 5–6 frames of brood. These bees were sure strange I thought—certainly nothing I had experienced before.

When the colony was closed, only some of the adult bees about our heads returned to their hive. By now we had to shout to be heard over the noise of their wings. Inspecting the two other colonies produced the same response, though they also were weak colonies. By the time we finished, which was rapidly since we wanted to close the colony rather than spend time inspecting it, we had a halo of bees flying around our heads and several stings embedded in our protective equipment.

We exited the apiary by walking through the brush further up the hillside, puffing great quantities of smoke about our heads and bodies. We walked a long distance (500–1000 m – 1640–3300 ft) but I still had bees flying about my head when we came back to the flat ground behind the school buildings. We went into a classroom at least 300 m (1000 ft) away from the apiary before we removed our gloves, coveralls and veil. A person outside, not with our group, was stung when he attempted to wave a bee away from his face.

Since it was a hot day it was a relief to get out of the heavy coveralls. I did not feel like the time had been very productive although we had looked at all three colonies. I felt smoked, very hot and sweaty. What had I learned? My impression at that moment was one of relief. Most of our time had been spent preparing to enter the apiary, and then exiting via the long, circuitous route back to the classroom. The three colonies were not strong, all being in a single Langstroth box. They had a lot of brood for their size but virtually no food reserves. We saw the queen in one colony and young brood in the other two.

After this encounter, I went on to David, in the western province of Chiriquí, where Africanized bees had only been reported in the area for one year. Bee colonies here were normal. Most of the beekeepers believed the swarms had passed on and their area would not be affected. The harvest in 1983 had been very decent and the large reserves I had witnessed in 1982 were mostly sold (some for exportation). A few of the beekeepers welcomed the arrival of the many swarms since they were easy to capture, though they noticed that some did not always stay in the box. Most beekeepers realized they needed to leave their capture hive in place for 3 days to a week before they tried to move the swarm into an apiary. Swarms moving into stacks of bee equipment were a common occurrence.

Changes at McClymonts

During the same 1984 trip, I revisited the McClymont apiary. Sam had died earlier in the year before I returned. I went with his grandson to a rural apiary and we tried to inspect a couple of the colonies but they were very defensive so we quit. I talked with Greta about sources for obtaining honey since they were not going to have a harvest. Beekeepers in western Panama still had honey to sell as they were only in the first season of Africanization.

Perhaps it is fitting that Sam died before he had to change many of his ways. His last year of beekeeping (1983) was his most difficult as Africanization was proceeding. His neighbors demanded he move those bee colonies we had looked at in 1982 out of the backyard after their dog and one child was stung. The bees had suddenly became agitated; Greta wondered if it was in response to the strong odor of disinfectant used to wash the patio floor. She said they had lost locations because of stinging incidents. They had to kill colonies that became unmanageable and had to abandon some sites.

The pace of Africanization had accelerated in 1984. Greta and Sam's grandsons could not find sites, and with Sam ill, they were not able to adequately manage the bees. Eventually they were reduced to one rural site but Greta worried about the neighbors' livestock. They had already paid for one visit by a veterinarian to treat some animals "their" bees had stung. Honey extraction was very difficult at the shed in their yard since "swarms" of bees would be attracted every time they had honey about.

I left, optimistic that Greta would find honey to supply her markets especially after what she endured initially to get her product started. I hoped the grandson would quickly learn and adjust to the new bee. Unfortunately none of this came to be. Miel Mañana labels no longer grace the supermarket stores or roadside stands and Greta and the family are no longer beekeepers or honey retailers. The McClymonts became one of many victims of Africanization.

Other First Experiences

Were my experiences unique? Here are two other descriptions of first experiences to share with you.

Mark Winston described his recollections of his first visit to a colony of Africanized bees, which was a small one. His description leaves little doubt that he[2] was looking at something different:

> "I arrived in French Guiana late one night in the summer of 1976, and the team went out early the next morning to see our first Africanized bees. When we opened up the first colony, I was still keyed up from travel and jet lag; my adrenalin was flowing partly from the excitement of beginning my research but also because I was fearful of a vicious attack from the supposed killers. To my surprise, the bees did not attack us, but seemed to be even more nervous and highstrung than I was. Although this was a small colony, the activity level inside was astounding, with worker bees running in waves across the comb and down through the hive. Even the queen, who in European colonies is slow and ponderous, was, in beekeeper terminology, 'runny'."

His recollection of a visit a year later to a beekeeper's apiary was more similar to my initial experience. He needed to go to the neighboring country of Surinam to visit an apiary since Dr. Taylor found only 15 colonies of honey bees in all of French Guiana before arrival of the Africanized bee. Mark writes:

---

[2]Note: The "team" reference was to graduate students from University of Kansas.

"My 1977 visit to an apiary in Surinam that had recently become Africanized was typical. A local schoolteacher maintained about twenty colonies in an isolated grove out in the country, a few hundred meters away from a number of small farms. He abandoned his colonies shortly after my visit, but at that time he was still keen on beekeeping and eager to convince me of the advantages of these bees. Even so, we parked about a half-kilometer from his beeyard, put on two layers of clothes under our bulky beesuit coveralls, and carefully secured our veils and gloves to leave the bees no room to enter. Then we lit the largest smokers I had ever seen, bellows-like instruments that burn burlap, old sheets, cardboard, dried cow patties, or whatever is available to generate smoke to pacify the bees. Only then did we approach his colonies, and I should have been warned by these elaborate precautions.

"Merely walking toward the colonies elicited a massive response on the part of the bees, so that the situation was out of control before we smoked and opened our first colony. Bees were everywhere, banging into our veils and helmets with such ferocity that we could barely hear each other and stinging through our layered clothing. It was a hot, humid day, and the combination of sweat, noise, and stings forced us to retreat after examining only a few colonies. The bees followed us all the way back to the car, and we had to keep our equipment on until we were far out of their stinging range. As we drove off, we could see the farmers swatting at bees and two of their cows were being stung; we had to stop and move the animals farther away to safety.

"Fortunately, no one was seriously hurt, but this incident was not unusual...."

Dave De Jong described his first experiences with Africanized bees 15 years ago in Brazil in a 1994 *Gleanings in Bee Culture* magazine article:

"My first experiences with Africanized bees began in 1980. I already had considerable experience working with European bee colonies in the U.S. As a bee inspector I had often opened and searched the brood nests of 100 or more bee colonies per day, from eight in the morning till late in the afternoon, in all kinds of weather. I took a lot of stings and never used gloves. For this reason the adaptation to Africanized bees was not so difficult. And I had the advantage of observing and working with Brazilian beekeepers and researchers with considerable experience. Nevertheless it was obvious that something was different. The bees repaid any mistake quickly and efficiently. What most impressed me were the nuclei. While large colonies in the U.S. were well able to defend themselves, we worked the nucs practically without smoke and frequently without a veil. The Africanized nuclei were quite different. Even a nuc of four or five frames had the capacity to send out hundreds of defenders in seconds. Obviously I learned to respect them. Later when I visited the U.S. and Mexico, I was surprised at how 'slow' the (European) bees were. After getting used to the Africanized bees, the Italian bees seemed to have something wrong with them."

Have you had a chance to open and inspect a colony of Africanized bees? I hope you get the opportunity—as a visitor in another apiary rather than with your own colonies.

✧✧✧✧✧✧✧✧✧

I once read a description of an initial encounter with the tropical jungle. Mary Batten described her trip to the Panamian jungle as a "journey into another biological dimension. It explodes the senses with a rush of colors, odors, sounds and life forms. It is a journey to eternal summer…a journey to the dazzlingly beautiful…it is a journey to our ancestral home." I, too, vividly recall my first visit to the jungle in Panama. I also recall, as if it were yesterday, my first encounter with Africanized bees in Panama. Both need to be experienced to be believed!

7

AFRICANIZATION OF PANAMA

The Africanized bee brought many changes to beekeeping in Panama. Within two years, the invading Africanized bee population completely replaced the European-race bees. Beekeepers experienced many problems trying to adjust. (Chapter 6)

A short visit to Panama in 1985 revealed additional changes to those I found in 1984. There was very little honey to harvest and beekeepers began to report large numbers of defensive colonies that were hard to manage. Many beekeepers had changed their apiary site or were actively looking for alternative sites. They all had to dress in heavy clothing for an apiary inspection

Beekeepers were now reluctant to invite me into their apiary to look at their colonies because of their concern that inspection might lead to a stinging situation. They were discovering that people and/or animals were getting stung after they opened colonies. Nearly all said they had eliminated one or more colonies; they no longer were enthusiastic about capturing swarms.

When I revisited Hermann Gnaegi in 1985 several changes were evident. I saw only three hives near the house; he had moved most of his colonies to one of his pasture areas. We talked about isolating apiaries and building fences so animals could not directly contact hives. He did not wish to don his equipment and inspect colonies while I was there—saying he was too busy. He had not been able to purchase honey for his export business; no one had honey to sell.

Hermann Gnaegi was one of the commercial beekeepers of Panama to give up bee culture. He still maintains a few colonies, but during a visit in 1989 he was asking a beekeeper, who accompanied me to his house, about buying honey to make his (still excellent) honey wine. He has continued his fruit and cashew nut export business but no longer deals in honey.

A tour with 24 U.S., Canadian and Mexican beekeepers (collectively owning 100,000 colonies) began a Central American tour in Panama in early 1986. Reports of that trip by Sue Cobey and Sarah Locke, and another by Troy Fore, documented changes Africanized bees brought to Panama. The tour visited the Panama Canal with Ernesto Cordovez, head of the Panama Canal Commission "Bee Swat" team responsible for controlling Africanized bees in the Canal Zone and on ships transiting the canal. Tour members also stopped in Penonome, Divisa, and also visited with Juan Malivern in western Panama.

Figure 7–1. Suburban Panama City apiary pre-Africanization. (Note bus stop at shed in background.)

Stinging Incidences

As I revisited beekeepers in 1984 and 1985, I began to learn of the human cost of Africanized bee colonization. Bee colonies, valuable livestock to the rural campesino, had become a dangerous liability!

Africanized bees, when disturbed, exit the hive in great numbers and search in the area around their home—in the case of the campesino, the clearing around the structure where the family lived and animals owned by the family were kept. With defensive Africanized bees, these animals, due to confinement or because they were tied, were stung hundreds even thousands of times. Some died from excessive numbers of stings.

People attacked by stinging bees usually seek to escape inside a house. Unfortunately, the family home in tropical areas is not always an adequate escape site. One of the differences with Africanized bees is that they will fly into dark places, especially if they continue to perceive the odor of their alarm chemical imbedded in clothing of a person or hide of an animal trying to escape.

Beekeepers told stories about stinging incidents with their bees. Such incidents often had severe economic consequences for a family.

Some lost livestock, or a family member was injured requiring a visit to the health clinic. Beekeepers' families were often the first to suffer an accident. Few beekeepers had alternative sites for relocating their hives so many colonies had to be killed by their owners. Some apiary sites were simply abandoned to be quickly overgrown by the jungle. Signs of neglect were everywhere.

When accidents occurred in city, town or suburb, the news media often reported the incidence. Radio and newspaper reports of the disturbance and consequences were often exaggerated and not always what really happened, particularly after the media arrived on the scene.

Media seek news and there is plenty of newsworthiness to a story about invading bees from Africa attacking innocent people, seemingly without provocation. USDA bee scientists Nunamaker and Stoner and Bill Wilson published articles illustrating that the press has not always been accurate with the real facts as regards the "killer bee." Unfavorable insect publicity is not confined to just honey bees. Moore, in a survey of 40 magazines, found 80% of articles "...unfavorable to insects." In the survey, honey bees were included in 17% of the articles but only 4% cited their beneficial aspects.

Panama's First Sting Death

When I revisited the McClymonts in Arraijan (Chapter 6), I had the opportunity to investigate a stinging incident that resulted in a human stinging death. Sam McClymont had passed away and Greta was trying to keep the business going with a grandson. I went with Greta to a neighboring town to talk with Pantaleon Rodriques, the beekeeper blamed for the accidental death. He had started keeping bees 6 to 7 years earlier with Sam's help.

The man stung to death was a 79-year-old who lived next to his daughter in a bamboo hut with a thatch roof. Every day this old man would walk to the neighborhood store to buy two cigarettes, all his daughter permitted him to smoke. The activity consumed most of his day. One day his trip ended tragically.

His path to buy cigarettes took him through a vacant lot with a single tree. For some reason bees nesting in the tree attacked him. Neighbors heard the man shouting with bees furiously stinging him. Several persons received stings trying to help cover him with a cloth. It was a couple of hours before a neighbor, covered in a blanket, was able to carry the old man to the health clinic. Bees were reportedly still stinging at this time.

The health technician said the man was still alive when brought to the clinic. However, he died during the night. Twenty-two other persons were treated for stings. The person that carried the old man to the clinic had to go to the hospital but he recovered. The clinic nurse estimated the old man who died had received over 500 stings.

That evening the tree where the bees lived was doused with gasoline and burned. Some members of the community tried to kill the remaining three colonies of Pantaleon, the local beekeeper who was blamed for the accident. Two of his colonies were destroyed but one survived The bees were very defensive when Greta and I visited Pantaleon to help transfer them to an intact hive body. Sam's grandson was to return that evening to take the colony to the McClymont apiary outside of town.

Nobody knows what agitated the bees. Maybe the man leaned against the tree; maybe kids were throwing stones at the bees; or maybe the bees were disturbed by a neighbor working on an old car in his yard. Pantaleon was being blamed because they were "his" bees. Bees had stung and killed some chickens of a neighbor sometime before the accident with the old man, so the apiary site was known.

Beekeepers may be held responsible, or blamed, regardless of what the law might say when such accidents occur. When bees become a source of irritation, the beekeeper often is forced to move the apiary or quickly modify the apiary to continue with bees at that site. In Panama, as elsewhere in the Americas, beekeepers usually pay the value of an animal killed in a stinging incident if their bees might have been responsible. Here the bees clearly were not the beekeeper's but he, too, was a victim.

Quebrada Bonita

In another visit to Panama, I went with José Vidal Gonzalez, Region 6 bee specialist with the Ministry of Agriculture (MIDA), to the town of Quebrada Bonita on a bee call. Although the town name means "beautiful little stream," the stream that went through this community of about 100 homes was badly despoiled. The store owner had called the MIDA office the previous day reporting that honey bees suddenly started to sting people, had killed a horse and were creating a problem in the community.

We talked to one neighbor who had swelling under his eye. The store owner who called MIDA said he had been stung repeatedly but there were no signs of stings. As we talked, the one horse became 3 killed—or maybe one killed and two injured. The number of persons sent to the health clinic was 10 or 30 (it really was only 2 or 3, with 30 more feeling they should have gone). The person most seriously stung was the owner of the horse that was killed, which at the time of the attack was tied to a tree branch.

We could only find a hundred or so disoriented bees flying around the door of a long-abandoned car frame in the trash-strewn stream ravine. A front-loader of Ministerio Obrero Público (MOP), the Public Works Ministry, (the road people), was at the top of this embankment. The story we pieced together suggested that the front-loader, pushing dirt over the embankment, had probably disturbed a feral Africanized bee colony nest ing in the car frame. The front-loader operator was long gone and nobody had seen him since the stingings began.

There were no bees left to remove. During the night neighbors had doused gasoline on the car frame and set it afire. The flames burned part of two trees and almost burned the power lines overhead. There were still enough bees left over that people rushed by that section of the road covering their heads with cloth or clothing. While we were there one school youngster was nearly hit by a car while hurrying along trying to avoid being stung. Before leaving we sprayed some soapy water on the car frame and assured the citizens the problem was resolved.

Many other stories could be told about Africanized bee stinging accidents or incidences. There have been more than one instance when I realized I had no control over Africanized bees. More smoke does not help. You cannot get the bees away from your veil or out of the location where they have found entry to sting. The number of bees about your head seems to continue to grow.

I would say that a least two times I was scared that I might not survive. Other times, I have been fortunate to have a plan for retreat from the bees. After a couple of such incidences, you look in advance and plan ahead for what you might do just in case there is an emergency. It is a scary, very weird sensation when you reach that point. I equate the feeling to when you are being shaken by an earthquake. It is so unique, so distinctive, that you will never forget it, yet find words inadequate to describe the sensation.

Is this any way to keep bees?

National Commission

A national commission (Comision Nacional para el Control y Manejo de la Abeja Africanizada en Panamá) was established as the Africanized bee arrived in Panama. This was developed partially at the initiative of Dr. Marshall Levin of the USDA who had visited Panama with Dr. Orley Taylor in 1978 to talk with agriculture and health officials about the Africanized bee. Membership on the commission included the Ministry of Agriculture (MIDA), representatives of other appropriate Ministries (Health, Education, Finances), the University of Panama, Smithsonian Tropical Research Institute (STRI), and beekeepers. The Canadian International Development Agency (CIDA) provided funding for this commission.

A major activity of the commission was to mount an intensive public education campaign about the Africanized bee directed toward beekeepers and the general public. Radio spots were developed and information was fed to newspapers. A poster, modeled after one used in Venezuela, was widely circulated. (Figure 7–2) It was headlined *"Do not attack the Africanized honey bee."* At the bottom it included the statement *"Increasing production of bees is necessary for the country."* The body of the poster had an attractive line drawing of a youngster about to throw a stick at a swarm (a hornet's nest?) with a rural school (flying the national flag) in the background.

There were two columns of information below the drawing. On the left was information about Africanized bees: *"The Africanized honey bee is with us,"* and *"The Africanized honey bee must be considered a resource for production of food."* On the left was information, such as: *"Prevent accidents which swarms can cause,"* and *"Don't work close to the bees if you don't have proper protective equipment"* and *"In case of emergency call the fire department and communicate with MIDA."*

Figure 7–2. Panama Commission bee poster.

The educational campaign is generally credited with helping keep stinging incidents in Panama to a low number. The first human fatality did not occur until March of 1984 (described above), almost two years after arrival of the invading population, in sharp contrast to the experience in Venezuela where 100 deaths occurred the first year of colonization.

At the 1984 Entomological Society of America meeting in San Antonio, TX, Clyde Stephens, entomologist for United Brands, western Panama (Chiriquí), cited the usefulness of the Commission program. He said that his company met with members of the commission and "got a lot of good information." This was used to educate their 6000+ employees resulting in "a minimum of panic in the banana zone."

Another commission activity was to educate beekeepers and to assist them in adapting to the new bee. The Ministry produced several mimeographs and attractive printed brochures on Africanized bees authored by Ramón Vanegas, then head of the MIDA bee program, Dr. Molino Pardo of Colombia and Dr. Mondolfi, an M.D. from Venezuela. Brochures contained useful recommendations on establishing a new apiary site, using individual hive stands, information on how to manipulate colonies, cautioned against working alone, suggested working bees after 3 P.M., and included the admonition to always use a smoker. The leaflets discussed the variability between the bees and their reactions when manipulated. They advised beekeepers to select and work with gentle colonies.

Along with beekeeper and public education, a series of training programs were initiated to train Ministry of Agriculture (MIDA) personnel and firemen in removal of swarms and feral Africanized bee colonies. MIDA personnel and firemen were supplied with appropriate protective equipment.

The fire companies (bomberos) soon were busy with bee removals. In David, for example, 87 swarms were removed from the city of 50,000 in the first two months of the dry season (January and February). During the first four years, firemen in Panama City were called upon to control 6400 bee colonies.

Panama Canal Response

There was a separate bee control program initiated in the Panama Canal Zone, carried out by the Panama Canal Commission and the U.S. Armed Forces. Part of the effort was development of a "Bee Swat" team, trained to respond immediately to reported sightings of bees and to eliminate swarms and feral nests in the Zone.

The information on bee/human interactions was systematically collected and recorded. In a 12-month period (April 1985–March 1986), 35 stinging accidents of work crews were reported. In a 48-month period, 1175 bee colonies and swarms were destroyed in the 50 km^2 (19 mi^2) Canal Zone area. The number of Africanized colonies found in the Canal Zone increased dramatically, though the number of swarms and number of colonies removed dropped in the last year of the study.

Table 7.1

Percentage increase (+) or decrease (-) relative to previous 12-month period for total number of swarms and nests controlled by Panama Control Commission teams

Apr–Mar (12 mo.)	Change (%)	Total no. swarms	Change (%)	Total no. nests	Change (%)	Total (swarms & nests)
1982–83	—	175	—	60	—	235
1983–84	+51	277	+58	78	+30	355
1984–85	–3	239	–14	105	+35	344
1985–86	–30	172	–28	69	–34	241

More than 70% of the 281 nests destroyed by the Bee Swat team were associated with artificial sites—32% were on or near homes, buildings or work facilities of the canal; 41% were located in water-meter boxes, utility manholes, hollow cement light poles, empty oil drums, pipes and air-conditioning ducts. Only 26% of the reported sites were in natural locations such as hollows in palm trees, rock piles, aerial termite nests and the like. In 48 months, swarms were destroyed on 48 ships along with an additional 243 from the dock and lock areas of the canal.

The U.S. military uses the Panama Canal area for jungle warfare training. A soldier stationed in the Canal Zone, Willenjohn, described a bee attack on three employees who were maintaining a jungle trail used to test soldier skills and equipment. The individual who jumped in water and covered his face with a mesh baseball-type cap received a large number of stings while immersed and more when he left the water 20 minutes later. The individual who initially disturbed the bees when he stepped over a log where they were nesting, quickly ran and escaped with few stings. The third individual sprayed the bees with insect repellent DEET® and ran; he received several stings. As a result, DEET insect repellent has become standard emergency equipment for soldiers and employees.

Africanized Bee Impact on Beekeeping

In 1982 when the Africanized bee arrived (Chapter 6), there were 338 beekeepers and over 18,000 managed colonies of European-race bees in Panama. Production average was approximately 8 gal/colony. Since 1982, the number of colonies has declined by over 80% and total production is down 90%. The number of beekeepers has dropped by over 50%, compared to the number before arrival of the Africanized bee. (Table 7–2)

In the Canal Zone, Africanization of gentle European-race bee colonies occurred rapidly as in previous tropical areas invaded by the

Africanized bee. Based on morphometric measurements, researchers Boreham and Roubik concluded that the entire bee population of Panama was Africanized within two years.

Table 7.2
Apicultural Statistics, Panama, 1974–1989

Year	No. beekeepers	No. beehives	Honey (gal.)	Avg. yield (lb/col.)
1974	—	12,000	28,000	27.6
1975	—	13,150	29,750	27.6
1976	—	14,550	63,700	52.8
1977	—	15,450	175,000	27.6
1978	—	16,816	111,500	79.2
1979	—	17,200	134,000	93.6
1980	338	17,982	143,856	96.0
1981	369	19,800	158,800	96.0
1982	369	21,806	174,448	96.0
1983	294	20,135	135,171	80.4
1984	243	16,242	45,739	33.6
1985	253	13,814	40,792	36.0
1986	210	8,037	32,258	27.6
1987	192	8,143	32,556	48.0
1988	—	—	—	—
1989	179	3,613	18,693	62.4

Changes

Many changes occurred in beekeeping in Panama following arrival of the Africanized bee. Honey bee colonies are extremely defensive and colonies cannot be examined without protective equipment. Clothing must be donned before entering the apiary and cannot be removed following inspection until well away from the bee colonies. This is a great discomfort, of course, in the hot, tropical climate of Panama.

Virtually all apiary sites have been changed, sometimes repeatedly, and isolation from people and animals is a must. Finding appropriate sites is a real challenge for beekeepers. Most beekeepers use the large volume (Brazilian) smoker with one individual puffing the smoker the entire time apiary work is being performed. Colonies are now mostly placed on single hive stands, one or more meters apart, rather than collectively, close together on a stand. (Figure 7–3)

Figure 7–3. Apiary in transition—2 colonies left isolated on stands while those in center are still grouped on stand.

Shaded sites are recommended and fewer colonies per apiary site are now the norm. There is still much evidence of colony neglect, lack of attention to unused equipment, inadequate termite protection or clearing of vegetation in the apiary. A general decline in apiary and colony maintenance is readily evident.

Management changes are evident as well. Colony inspections are more rapid and less frequent. Beekeepers now routinely feed their colonies to try to reduce the amount of absconding. They do not need to feed to stimulate but rather to try to keep the colony from absconding. Most feed dry sugar since there is abundant moisture during the rainy season and some bee flight is possible.

The Africanized bee is not as readily manageable as European-race bees. Beekeepers report that labor costs are higher with Africanized bees. Beekeepers use a larger work force and enter the apiary later in the day, manipulate colonies as quickly as possible and then exit the area, staying away from the apiary for at least a couple of days after an inspection.

There are clouds of bees around the veil and on protective clothing following colony inspection. It takes time to reduce this population upon leaving the apiary site. It is important not to bring these bees toward a residence or to areas with animals since the bees are inclined to sting.

Beekeepers report that protective equipment slows the bee inspection process and reduces work efficiency due to the bulkiness and discomfort of such attire in the tropical climate.

Beekeepers find Africanized bees somewhat different from their experiences with European-race bees before colonization. Colonies expand rapidly once there are nectar/pollen reserves available when the wet season ends in November/December. Colonies seem to "explode" overnight with brood. They begin to start rearing queens and quickly are ready to swarm. There are two annual swarming peaks, one in January–March and another in June. Along the much rainier Atlantic coast [3500 mm (140 in), compared to 1400 mm (55 in) along the Pacific], swarms are larger (megaswarms) and less predictable. Without swarm control, colonies are not as productive.

Before Africanized bees moved into Panama, beekeepers reported no feral bee populations. Swarming was not common and swarms were usually encountered only within the immediate vicinity of an apiary. Now swarming is common and feral colonies abundant. Beekeepers report that swarms readily move into stacks of empty equipment. These swarms are seldom used as a means of colony increase, however, since they frequently abscond. Beekeepers report that colonies started from swarms usually result in a highly defensive colony once the population increases in size.

Experienced beekeepers report yields of properly managed colonies as "similar to those formerly expected from European-race bees." Statistics (Table 7–2) do not support this simply because not all beekeepers have the necessary management skills to prevent swarming/absconding or take the time to intensively manage their colonies as needed with the Africanized bee. Beekeepers that do manage colonies skillfully report harvesting nearly the same amount of honey compared to pre-Africanization days. The recent arrival and spread of Varroa mites represents a new challenge to Panamanian beekeepers.

According to many beekeepers, honey sales do not seem to be adversely affected by the bad publicity the Africanized bees received. Removal of honey-filled frames is a challenge. Panama, with approximately 2.5 million inhabitants, is not a large honey-consuming country. There is too little honey to meet in-country demand.

There are no functioning government programs in Panama involving selection of gentler stock and no regular requeening with European stock as in neighboring Costa Rica (Chapter 8). IDIAP, the agricultural research agency of the government, initiated a program to gather and evaluate queens from producers in Chiriquí and the Central Provinces with the intention of distributing daughter queens from the most productive and manageable colonies. MIDA was to evaluate F_1 daughters reared from the queens. The program terminated after two years when highly defensive colonies (Figure 7–4) and heavy absconding led those in charge to conclude that the program would not succeed.

Figure 7–4. Highly defensive IDIAP colony used for queen rearing.

Higher elevation areas that might support large populations of gentler bees to establish queen mating stations are not readily accessible in Panama. Juan Malivern's program to rear queens in the highlands has not continued and a program to rear F_1 queens at Divisa terminated due to lack of customers. Some large islands in the Bay of Panama, such as a large prison island off Chiriquí Province, offer possible isolated breeding sites but there are no funds for a government-run program. No individual beekeeper has yet developed the business of rearing queens, in part because the industry is not very large.

Most successful beekeepers, such as Juan Malivern, have diversified by selling pollen and royal jelly in addition to honey. Dave Roubik and associates, from their base at STRI (Smithsonian Tropical Research Institute) in the former Panama Canal Zone, studied pollen collection cycles with hives in the Panama Canal and Chiriquí Province areas. This basic information has greatly assisted beekeepers interested in producing and harvesting this product. Beekeepers who sell pollen and royal jelly report that there is a good market for these products.

Surviving with Africanized Bees

One Panamanian beekeeper has managed to continue keeping European bees. He keeps bees on isolated islands such as San José Island. This island is part of a large number of Pearl Islands in the Bay of Panama. It is a large, flat island, approximately 5000 hectares (12,355 acres) in size. It is located 68 km (42 miles) from Punta Brava on the Darien mainland; the next closest island is 32 km (20 mi) away. Rey Island, the largest Island in the group, is another 18 km (11 mi) further east. Rey Island is 38 km (24 mi) from Punta Pedernales on the mainland.

San José Island is headquarters for George Novey, one of the few remaining commercial beekeepers in the Republic of Panama. His bee operation seeks to maintain European bee stock. He has 450 colonies in 15 apiaries of about 30 hives each; he averages nearly 85 lbs of honey per year. To maintain European bees, he regularly requeens his colonies with queens flown in from the U.S. Any defensive colonies are eliminated immediately and any swarms that reach the island are killed

Although his apiaries are on islands many miles from the mainland, Africanized bee swarms island-hop. The first Africanized bee swarm arrived on San José Island in April 1984. Novey experiences approximately 5% Africanization of his operation each year. Still he persists and is the largest honey producer in Panama, though securing queens and eliminating bees means beekeeping has become more costly and less profitable. Still, in comparison with the mainland, his problems are fewer.

Juan Malivern (Figure 6–5, page 71) remains one of the largest beekeepers in Panama, though his wife Estrellita is the beekeeper these days as Juan has returned to the bank business. To survive he has no choice but to manage Africanized honey bees. Survival has necessitated many adjustments and changes in his operation.

Juan is among a minority of beekeepers able to adjust and change management to the new bee. Most beekeepers choose not to continue beekeeping. It has been my impression that it is far easier to train individuals new to bee culture how to manage Africanized bees than it has been to retrain beekeepers familiar with European-race bees.

In 1986, three years after Africanization, a U.S. beekeepers' tour spent two days visiting with Juan seeing his business and inspecting Africanized bees in his apiaries. Juan's bees were described as "extremely aggressive." In one instance, one experienced beekeeper from the tour joined Juan and his crew in opening and inspecting colonies in the apiary while the others stayed a suitable distance away to observe. As happens with this unpredictable bee, one colony in the apiary became very defensive and chased the tour beekeepers "through the tall grass and down the road in an attempt to get rid of following bees" even though the beekeepers never entered the actual apiary site. Juan lost that site soon thereafter when people living nearby poisoned the colonies because they were continually being stung.

Juan no longer has hundreds of bee colonies—apiary sites are too hard to find. Colonies must still be managed and Juan believes intensive management is needed to produce good honey crops. Africanized bees do produce honey, and production in good years approximates yields realized with European bees before Africanization. But there are bad years when production is low, even with intensive management. Malivern apiaries now profits from bee-collected pollen more than honey sales.

Juan feels intensively working bee colonies helps him find and eliminate the worst-behaved bees and more hive visits conribute to better behavior. During the dry season of December-March, he harvests honey-filled frames 3 to 4 times. He has switched to benzaldehyde to repel bees since brushing results in too many stings in the apiary and possible trouble to people living in the vicinity of the apiary.

Juan feels the abundant swarms are a resource that are potentially exploitable. At present, he only traps the accidental swarm, mostly those that show up at his David extracting facility. Once swarms have sealed brood, they can be moved and used to combine with colonies needing boosting or requeening. Swarms can be trapped in a box baited with lemon grass. Capturing swarms is unproductive because too many quickly abandon the hive. Colonies can be moved, but care must be taken to close all entrances. The bees are quiet on the combs when being moved.

Juan remains confident of the future. Demand for honey is strong and, although competition for nectar sources is great with feral colonies everywhere, there are few competing beekeepers left in the marketplace. He feels he has no choice of what bee to use since the cost and time of trying to requeen each colony is prohibitive. His biggest complaint: "When African bees come to a country, your worst enemies are the government, people and newspapers. The African[ized] bee by itself is no problem."

Juan Malivern and George Novey survive with Africanized bees!

✧✧✧✧✧✧✧✧✧

Africanization has changed beekeeping and very negatively impacted beekeeping in Panama. The industry has yet to regain its former prominence. Human/bee interactions have not led to elevated numbers of sting deaths but many citizens know about and have had negative experiences with Africanized bees. Surviving beekeepers have had to make many adjustments to stay in business. Lack of government support has resulted in a slow recovery and a long-lasting negative impact of Africanized bee colonization on beekeepers and beekeeping in Panama. Prospects are not good for immediate improvement.

8

Beekeeping in Costa Rica

Costa Rica, like neighboring Panama to the east, is a mountainous, narrow country on the isthmus of Central America. It is 120–240 km (75–150 miles) wide and 442 km (275 miles) long. As in Panama, a Central Mountain chain divides the country into Pacific and Atlantic sides. Of the nearly 4 million inhabitants 60% reside in a large Central Valley at mid-elevation (914–1676 m – 3000–5500 ft) where the climate is delightful. Tourist brochures describe the climate as "forever spring"!

The season in Costa Rica is wet-dry throughout with tropical conditions along the Pacific somewhat different from those in the Central Valley, where it is more spring-like year round. There is very little beekeeping along the rainy Atlantic side and few colonies are found at higher elevations in the mountains (which reach to 3840 m – 12,600 ft) dividing the country.

Beekeeping in Costa Rica, as in neighboring Panama, developed relatively recently although honey bees were introduced into the country prior to 1830. Beekeeping initially developed in the coffee-growing area of the Central Valley.

In the 1970s, German (GTZ) development funds improved training of beekeepers, established a queen rearing facility at Turrialba and led to expansion of the bee industry. Beekeeping extension, dating to the 1930s, was enlarged. The tropical Pacific coastal provinces, Puntarenas and Guanacaste, replaced the Central Valley as the major honey producing region of Costa Rica. By the time the Africanized bee arrived in 1983, there were approximately 1000 beekeepers managing an estimated 30,000 European-race (primarily Italian) colonies in Costa Rica.

Invasion of Costa Rica

Arrival of Africanized bees was confirmed in February 1983 in a feral colony 65 km (40 miles) from the Panamian border. As elsewhere, the Africanized bee spread rapidly. Africanized bees were present in a National Park near the northern border with Nicaragua within the same year, a distance of 400 km (250 miles). Spread of the Africanized bee along the wetter Atlantic side was slower and of less consequence.

The beekeepers of Costa Rica, as those earlier in Panama, were initially convinced they could handle Africanized bees. Dr. William Ramirez, currently Professor of Entomology at the University of Costa Rica and beekeeper with 35 years of experience, says "most...beekeepers did not take care of the bees" when Africanization occurred. The "subsequent drop of the national honey production down to 40% or less" was due to failure to change according to Professor Ramirez. Numbers of managed and harvested bee colonies were "dramatically reduced because beekeepers did not adapt."

At the time of Africanized bee arrival, the Costa Rican Ministry of Agriculture (MAG) attempted to establish a queen-rearing apiary at 2200 m (7200 ft) elevation on the slopes of Poas volcano, where there was a small number of European-descendent settlers who kept European bees. The idea was to conserve European bee stock so it could be used to breed pure European queens to distribute to beekeepers to prevent Africanization of their colonies. Unfortunately the project never developed due to funding constraints and politics.

Before the arrival of the Africanized bee, queen rearing and requeening of bee colonies was not widely practiced in Costa Rica. Previous attempts by several individuals to rear and sell queens, stimulated by the German development project, failed because too few Costa Rican beekeepers were interested in buying queens. Since Africanization, queen rearing has become an important part of beekeeping in Costa Rica and a profitable business for a few beekeepers.

Arturo Salazar

Arturo Salazar is one of Costa Rica's queen breeders. He lives in Las Juntas, which is in the Pacific lowlands along the gulf of Nicoya in Puntarenas Province. Arturo sells European-race virgin queens raised from queen mothers he buys from Hawaii or San Andreas Island. The latter is an island, 50 km (30 miles) off the Caribbean coast of Costa Rica, that has not been invaded by Africanized bees. It is a popular weekend retreat for Ticos (as Costa Ricans are called) but perhaps better known as a stop on the "cocaine highway" from South America to the U.S.

Although Arturo maintains drone-source colonies, all the feral colonies are Africanized in his region so virgin queen matings are with Africanized drones. The hybrid mixture of European and Africanized bees (called F_1) results in colonies that are less defensive compared to the feral Africanized bees. Such colonies can be manipulated and are generally productive, at least in the central highlands.

When I visited Arturo in 1989, he was asking about the mite situation in the U.S. He said he preferred queens from the U.S. but it was illegal to import them as Costa Rica did not then have Varroa mites. The queens he could secure from Hawaii did not do as well as he wished.

I revisited Arturo in 1994. Though his production in January at the time of my visit was still low, Arturo had doubled his production from my visit 5 years earlier. He was selecting bees from among the Africanized bee population. The gentle nature of his colonies was evident as Arturo used an ordinary veil during colony inspections and did not use gloves.

Arturo had not changed his apiary location which was nearly surrounded by homes on the edge of the small town of Las Juntas. One thing in his favor was that the apiary site was enclosed by a vegetation barrier. At the conclusion of our inspection it was possible to remove veils and bee suits to take a photo with Arturo before exiting the apiary site. This would never be possible in Panama.

Defensive Bees

During a four-month stay in Costa Rica during 1990, I organized a visit to Arturo's apiary for Dr. Floyd Horn, top research administrator for USDA. Dr. Horn was in Costa Rica to visit the screw worm project in Cañas, in the lowlands near where Arturo has his queen-rearing business. Dr. Horn is not an entomologist so his visit in Cañas presented a great opportunity to introduce him to Africanized bees.

In preparation for Africanized bees, we suited up before entering Arturo's queen yard. Arturo showed us his breeder colonies and explained the process of raising F_1s. We spent 30–45 minutes examining several colonies and mating nucs in the process. Arturo did all this inspection without gloves. I was beginning to think Dr. Horn would go back to the U.S. with a false impression of Africanized bees.

I repeated my request to Arturo that I wanted Dr. Horn to see Africanized bees. He said we would but he wanted to wait so it would be the last colony inspected. He was very proud of his breeder queens and his queen-rearing business and he wanted to be sure Dr. Horn understood the effort it took to produce quality queens. Although a relatively new queen breeder, his family had kept bees for over 40 years. Arturo knew his business and wanted us to appreciate what he knew.

At the edge of the apiary were 10 colonies Arturo had just purchased from another beekeeper. (Figure 8–1) They still had their moving slats holding the boxes together. He had not yet requeened them, which was his intention. As we approached these colonies he made sure our bee equipment was bee tight and he checked his own, donning rubber gloves he carried in his pocket.

The first colony we inspected responded in classic defensive fashion. The hive was well propolized and it was impossible to enter without jarring the colony. As Arturo removed the moving slats and pried the top, the bees exited through our smoke screen as if it were not there. They were all around us, trying to get inside our clothing, popping at the veil front, and attacking the dark camera one person was carrying. Arturo pulled a frame and bees ran off the comb to beard at the bottom edge almost immediately. Many bees were in the air and we literally had to shout to be heard above the noise of flying and fanning bees.

Figure 8–1. Floyd Horn about to experience defensive Africanized bees in Arturo Salazar's apiary.

What had been a pleasant apiary visit turned into an ugly situation. We no longer had control over the colony. A second frame was pried free of propolis and lifted out of the hive body. More bees exited and flew about us. Arturo did not wish to go further; we also had had enough. By now everyone had 100 or more bees flying about our heads. The colony was closed; we heavily smoked our bee suits and walked to the woods adjacent to the apiary. We spent 15 minutes or more walking away to lose the majority of the flying bees before turning to walk back to our vehicles.

Dr. Horn was suitably impressed with his experience of classic Africanized bees!

Visit to the Perez Brothers

It was pure coincidence that the most defensive colony the group of 24 visiting U.S. beekeepers witnessed in a 10-day tour I guided in Costa Rica four years later was in Las Juntas. After visiting Arturo, our tour visited the Perez brothers, commercial beekeepers who are honey producers and also rent some of their 600 colonies for pollination of melons.

We toured their modern extracting facility and saw their woodworking shop. They make most of their own equipment, preferring to use native woods like Pochote, which they harvest and dry themselves. Our group had a lot of questions so we spent most of the afternoon with them. Then late in the day we prepared to visit an apiary that was at the edge of a farm that had horses and cattle, common in that region. None of the Perez colonies were in town or near homes.

It was too dark to take photos without a flash by the time we arrived in the apiary. The colonies were in a grove of trees and it was dark enough so it was hard to see where we were walking once we left the open pasture. Miguel Perez opened one colony (after smoking it) and the bees exploded from the crack in the cover and immediately began to sting. Ray Hicks, a California beekeeper, was stung several times through blue jeans at the level of the hive opening as he took a photo. (Figure 8–2) All Miguel did was remove the hive cover; no frames were removed.

Figure 8–2. Highly defensive Perez brothers' colony. (R. Hicks photo)

What was truly impressive with this colony was the persistence of the bees that exited the colony in the short time it was open. We left the wooded site and returned to the open field. By then it was too dark to see faces, only the outline of a person, yet bees persisted in flying about our veils and bodies. The horses and livestock had long since disappeared.

Our beekeeper Miguel acknowledged that it was late in the day to open a colony though they try to do most bee inspections in the late afternoon. He said he has to be sure it is completely dark before he begins to load colonies for pollination rental. He apologized and said his colonies did not normally react so explosively since he regularly requeened. Such defensiveness "would not be tolerated by the growers" who rented his bees for pollination he said.

Stinging Accidents in Costa Rica

Although stinging incidences like our tour experienced in the Perez apiary are common for beekeepers, there is no evidence Costa Rica has experienced a large number of human stinging deaths from honey bees since arrival of Africanized bees. The English-language weekly newspaper *Tico Times* on April 5, 1991, reported Guillermo Guiterrez to be the 12th sting death over an eight-year period. It went on to add that "more than 500 persons" have required treatment for stings in the same time period. Greg Flakus, who worked as a Voice of America correspondent in Costa Rica for several years, supports the same impression in his book *Living* with *Killer Bees*. As in Panama, a public informational/educational campaign deserves at least some of the credit for this low incidence of accidents as it helped inform the public how to behave around honey bees.

One of the more widely-reported stinging incidents involving Africanized honey bees occurred in July 1986. A University of Miami botany graduate student, Inn Siang Ooi, originally from Malaysia, was sampling plants in a remote area of one of Costa Rica's jungled national parks as part of a course from the Organization of Tropical Studies (OAS). He apparently disturbed a feral colony of Africanized bees on a rock outcropping and they began to sting him. He then fell or attempted to hide in a crevice but the bees continued to attack. Other students tried to rescue him but they, too, were stung repeatedly and had to retreat. It was after dark before they were finally able to reach Ooi but he was already dead. He was stung an estimated 8000 times. Three other students had to get medical attention for their stings, though they were all okay.

Was the reaction in Miguel's bees due to inappropriate inspection conditions? What preceded the tragic accident of Ooi? Beekeepers know there are some conditions when Africanized bees are more defensive than others. Colony inspections can be a trigger leading to a stinging incident. Beekeepers in Costa Rica report frequent accidents with animals. Virtually all lowland beekeepers have had to change apiary locations due to neighbor concerns or complaints.

Beekeepers Tour Costa Rica

The 1994 beekeepers' tour I guided was not the first to visit Costa Rica. Sue Cobey and Sarah Locke, along with Troy Fore, both described a December 1985 beekeeping visit in Panama, Costa Rica and Mexico, organized by Troy and led by Orley Taylor. After visiting the Panama Canal "Bee Swat" team and Juan Malivern in Panama, the tour joined Marla Spivak, a PhD student of Dr. Taylor, in San Isidro, Costa Rica, where she was studying the Africanization process, then only two years post-colonization.

Fore said several tour members received stings and "saw plenty [of] action" in San Isidro. Spivak's research documented a rapid turnover of the European-race population and "a rapid establishment of a feral

Africanized population" at this low-elevation site. The group visited the apiary of teacher Canilo Quezada Rivera. He had moved his apiary to a remote site out of town. The colonies had not been opened for two months when Dr. Taylor did so for the group to "experience the explosive defensive reaction" of Africanized bees.

The 1985 tour also visited the Central Valley. One beekeeper visited was Frederico Alvarado, a 600-colony beekeeper in the coffee-growing region of the valley. Frederico told about capturing a large number of swarms, something unusual for him because swarming was very uncommon in Costa Rica with European bees. He showed the group "noticeable" behavioral differences between Africanized swarms he had hived and his European-race colonies. At the time, the Central Valley had been invaded by Africanized bees for only one dry season and beekeepers were still learning about them.

One key to profitable beekeeping with Africanized bees Frederico had already learned was to feed Africanized bee colonies to reduce their absconding. Africanized colonies needed to be fed after the rainy season began, not to stimulate the colony at the beginning of the dry season as was the case with European-race bees. It was clear that Frederico was a perceptive beekeeper who was making adjustments to the different bee; less adaptive beekeepers quit beekeeping with establishment of the new bee.

1994 Tour

My 1994 beekeeping tour had the chance to visit apiaries of beekeeper Max Gonzalez at both lowland and Central Valley sites. In Coylito, Guanacaste Province, only a few miles inland from the Pacific Ocean, we saw the more defensive nature of Africanized bees typical of Costa Rican lowlands when we visited Max's former mating yard.

When we arrived, only a few colonies remained at the site; most had recently been moved to melon pollination rental locations in the same region. When we opened one of the colonies, a large number of workers flew out and proceeded to sting coveralls and pop at veils. One member of the group, Harold Libermann, was taking photos and drew in excess of 200 workers that vigorously stung his black camera and case. After closing the colony our group had to walk a long, circuitous route through a wooded area to return to the bus. We still had bees flying about and needed to board the bus to remove veils and coveralls; several bees pursued us into the bus itself.

The behavior of bees flying about the head, persisting in stinging attempts, is a consistent factor with Africanized bees. Bees attacking a dark camera, dark socks, or dark veils covering the face is common defensive behavior. It adds considerable time to colony inspections when beekeepers have to lose the halo of bees after colony manipulations. It meant additional time to our tour to avoid bringing the bees back to the bus or to a home where our hosts would frequently offer us refreshments after our apiary visit.

Later on our tour, we visited a queen rearing apiary of Max Gonzalez in a coffee plantation in the Central Valley. In contrast to the bees we saw at Coylito, the Africanized bees in the coffee plantation were very manageable. Most tour members used only their veils and no gloves, yet the bees were of the same stock that Max used throughout his 500-colony operation. We did not have the cloud of bees around our heads nor see the defensive reaction experienced just a few days earlier in the lowlands.

One adjustment Max has had to make to Africanization was to switch his queen rearing operation from the lowland site in Guanacaste to this Central Valley site. The difference he and others felt important was that fewer feral colonies (and hence African drones) existed at the higher elevations thus Africanized bees were not so strongly favored.

Studying the Process of Africanization

Soon after arrival of the Africanized bee population, Marla Spivak began to document the process of Africanization in Costa Rica. She established apiaries at three different locations and elevations. Her lowland site was at San Isidro 700 m (2300 ft) in the tropical Valle del General, 65 km (40 miles) from the Panamian border.

In her study, Dr. Spivak used behavior and cell size measurements to determine the identity of colonies. For the behavior measure, Dr. Spivak and an assistant donned full protective bee equipment and slowly and carefully inspected and manipulated individual colonies periodically over several months.

Based on a rating scale, bee colonies were classified as (1) strongly Africanized: colony manipulation resulted in more than 10 stings, bees not calm on combs, queens hard to see, brood area extensive; (2) intermediate: zero to 10 stings, bees controllable with smoke; or (3) strongly Europeanized: two or fewer stings, bees remained on comb, queens easily located, brood not covering entire comb face. Three measurements of 10 adjacent worker cells from natural comb were also made. The colony was called Africanized if the mean of three measurements of the cells was 4.6–5.0 mm and European if 5.2–5.6 mm.

In San Isidro, Dr. Spivak was able to intensely monitor 18 colonies. Five (28%) were judged strongly Africanized, one strongly Europeanized and the remaining 12 were intermediate. In interviews with beekeepers, Dr. Spivak found that beekeepers quickly learned that inspection techniques formerly used for European-race colonies would result in extremely defensive behavior from one or more colonies that "made it undesirable or impossible to inspect other colonies in the apiary."

When Dr. Spivak began her studies, only one year following colonization, "virtually no one was producing honey" in San Isidro. Only seven of 86 beekeepers continued to manage their bee colonies in the region which, prior to Africanization, had an estimated 2500 colonies. Reports of "numerous stinging incidents of both people and livestock" were taken by Dr. Spivak.

One beekeeper, Ronald Montenegro, persisted by adjusting his management. In 1986, Ronald harvested an average of 40 kg (90 lb) from 90 colonies. He was able to do so by: (1) changing apiary sites to more isolated locations; (2) putting colonies on individual hive stands; (3) using good, sturdy protective equipment; (4) feeding sugar during dearth periods; (5) managing colonies to reduce swarming, and (6) identifying those colonies most defensive and inspecting them last.

Central Valley Africanization

An intermediate elevation site in Dr. Spivak's study was an apiary located in the Central Valley in the capital of San José at 1100 m (3600 ft). This is a highly urban area where nearly 50% of the population of Costa Rica lives. It has a large mixture of flowering plants, including coffee which is cultivated up to 1500 m (5000 ft) elevation. Average yields in this region, based on interviews conducted by Robert Kent in 1979, were 10–15 kg/colony (22–33 lbs), though more experienced beekeepers near coffee plantations produced between 30–40 kg/colony (65–90 lbs).

Swarms of Africanized bees were just arriving in the Central Valley when Dr. Spivak began her studies in 1984. The first major noticeable difference reported by beekeepers was an increase in number of swarms. In the two years Spivak conducted her studies (1984 to 1986), 12 of 29 colonies (40%) were classified as strongly Africanized; only one was judged strongly European at this Central Valley site.

I happened to visit Costa Rica in 1984, one year after Africanization. The beekeepers I talked to did not believe Africanized bees were present in their apiaries. They recognized that there were more swarms which previously was an uncommon event with European bees. Dr. William Ramirez of the University of Costa Rica hived a swarm that was "more nervous and runny," so he labeled it an Africanized bee colony. In Costa Rica I looked at colonies using just a veil for sting protection, in sharp contrast to a trip two months later to Panama where I encountered my first highly defensive Africanized bee colonies. (Chapter 6)

By the next season, 1985, differences in the bee population were noticeable to Costa Rican beekeepers. Swarms became very common and feral colonies were reported throughout the region. Ministry of Agriculture (MAG) beekeeping personnel and firemen were removing numerous swarms. A public education campaign, similar to the National Commission in Panama (Chapter 7), was in full force to educate the general population to avoid Africanized bees.

The pace of Africanization accelerated by 1986. Several reports of visitors to Costa Rica documented the measures surviving beekeepers adopted to remain successful. Dr. Spivak believes the attentive beekeepers were able to maintain the intermediate type of Africanized bee (as per her criteria) during the colonization phase by eliminating or requeening the more defensive colonies, using protective clothing and giving more attention to colonies.

Honey yields dropped by 25–50% during colonization of the Central Valley. However in a side-by-side comparison study conducted by Dr. Spivak of both Africanized and European colonies placed in a large coffee plantation in March–April 1986, three years after Africanized bee arrival in the region, no significant differences in honey production were found between colonies of the two races. The average yield of 41.6 kg/colony (92 lbs) in the study was not very different from earlier reported yields.

Table 8.1
Amount of honey harvested from colonies grouped by their behavioral characteristics, Costa Rica Mar 5–May 29, 1986.

Colony Type	Number	Honey Harvested (kg)
European	5	30.4 ± 14.6
Intermediate	6	50.3 ± 15.6
Africanized	4	42.5 ± 15.2

In the honey production study, colonies were not pre-selected for the experiment. During the course of the honey flow, the colonies were examined five times and rated European, intermediate or Africanized based on their behavior. Two colonies judged to be Africanized did not produce honey—one swarmed and one absconded so they were not included in yield calculations.

The study noted a greater number of management difficulties with the Africanized bee colonies.

High Elevation Studies

The third Africanization study site of Dr. Spivak was at 2000 m (6600 ft). She worked in an apiary of rustic hives, colonies without movable frames, which were robbed by cutting out the combs. At this elevation, swarms were more common, especially from the rustic hives which were 20–50% smaller than standard Langstroth boxes. Frosts occur at nights for part of the year at this elevation and rainfall is abundant. During the dry season, a warm, intense sun results in pleasant mid-70°F temperatures. Native vegetation has been displaced by dairy cows, strawberry culture and an expanding export flower and fern cultivation.

In 1985 Dr.. Spivak detected the first Africanized bees at this elevation (labeled as intermediate from her cell-size measurements and "irritable" behavior) at the beginning of the rainy season. All other colonies were rated intermediate or European. During the study, the beekeeper killed one colony and moved his apiary site, initially adjacent to the house, to a more distant location because livestock were being stung. He abandoned his apiary before the study concluded.

Comparison of Elevational Differences

To further expand her study and better define the Africanization process at higher elevations, Dr. Spivak established European-race and Africanized bee colonies at 900 m (3000 ft) in Alejuela at the University experimental farm in the Central Valley and at 1800, 2200 and 2800 m (5900, 7200, and 9200 ft respectively) on one of the volcanic mountains encircling the Central Valley. All sites were human-disturbed locations.

The seasonal cycle at the three higher elevation sites was similar. The dry season is the major honey-producing season; it begins in December and extends until heavy rains in August. The rainy season lasts four months, August–November. The mean minimum temperature at the 900 m (3000 ft) site, approximately 20°C (68°F), was the maximum mean at the highest elevation. The minimum mean at the highest elevation was between 0°C to 5°C (32–40°F). Rainfall was heavier at the higher elevation but the seasonal pattern was similar.

Forty-four of the original 62 colonies in the study (71%) survived the first year of the study. Colonies that died did so due to starvation, absconding, failure to replace queens, and the relatively poor conditions during the rainy season. Losses were heavier among European than Africanized colonies. Therefore, the second year of the study included only a comparison between the 900 and 2800 m sites after colony positions were reversed, i.e. the survivors at 2800 m (9200 ft) were moved to the 900 m (3000 ft) site and vice versa. In the second season, colony losses were similar at both locations.

Dr. Spivak found that feeding the European colonies during the rainy season was more important for their survival than feeding of Africanized colonies. European colonies dwindled in strength while Africanized colonies tended to abscond. Swarming during the dry season occurred three times as frequently in Africanized bees as in the European-race colonies. Africanized colony peak swarming was a full month before the European colony peak. Rates of swarming and absconding combined was very different for the two races—36 instances in Africanized bees versus 10 for European

It was significant that swarms did occur in the European bees in this study. Previous to Africanization, beekeepers uniformly reported little swarming from their hives. Swarming of European bees occurred at both the 900 and 2800 m (3000 and 9200 ft) elevations in approximately equal numbers (5 versus 4 swarms). Overall the Africanized colony swarming rate (77%) was considerably greater than in European colonies (38% of European colonies swarmed).

For her studies, Dr. Spivak maintained her colonies in a standard Langstroth box with 10 frames. In all parameters measured, Dr. Spivak found colony differences "most pronounced" at the lowest elevation site—900 m (3000 ft). She did note that all the bee colonies at 2800 m (9200 ft) "were relatively docile" and "the extreme defensive behavior characteristics of the Africanized colonies in the beginning was not noted towards the end of the experiment" at the high elevation sites.

In a 1996 report, Dr Spivak recalled that "the behavior of European colonies was relatively consistent over all elevations; they were relatively gentle and did not sting often or persistently." Africanized colony behavior "varied from visit to visit" and they "were variable within the apiaries." Spivak concluded from her study that Africanized bees have a wide range of characteristics, not a "well-defined, invariable phenotype."

High-elevation studies help provide clues as to what might be the eventual situation as regards Africanized bees in the U.S. Although the bees have readily colonized the tropical areas of the Americas, they dispersed upward in elevation less readily. Evidence exists from Spivak's studies in Costa Rica, Villa's work in Colombia and from my survey in Bolivia (Chapter 9) that the feral bee population is less common and defensive behavior less dominant at higher elevation. The density of feral colonies is lower and rates of swarming and absconding are lower.

Spivak's studies in Costa Rica identified several factors such as fewer resources, less successful queen replacement, colony dwindling and colder temperatures as possible reasons for reduced presence at higher elevations. She found fewer differences between Africanized and European race bees at higher elevations (2800 m – 9200 ft) versus greater differences at a lower elevation (900 m – 3000 ft). She did not consider either bee as ideally suited for the higher elevation conditions since mortality was high and productivity was low. However, it is very clear that Africanized bees will migrate to higher elevations and swarm/abscond at such locations.

Changes in Costa Rica

The 1994 tour I hosted visited large and small beekeepers in both the Central Valley and in lowlands all the way to the Pacific shores of Guanacaste Province close to the border with Nicaragua. Virtually all the remaining beekeepers in Costa Rica, and the new beekeepers who have started since Africanization, attempt to maintain hybrid (F_1) stock. In the lowlands, this bee competes well with the Africanized feral population. It is also a good bee for the Central Valley which again has become the major beekeeping region of Costa Rica. All beekeepers voice the same opinion that Africanized bees are easier to manage at the higher elevations than at lower ones.

In the lowlands of Costa Rica, there are now fewer beekeepers and fewer managed colonies. Typical is the beekeepers' cooperative at Hojancha on the Nicoya Peninsula. Some 25 beekeepers maintaining about

1500 hives still remain, less than one-half the number before Africanization. Agriculture in the area is changing with an increase in coffee plantations so perhaps more individuals will start bee colonies. Peace Corps volunteers have been assisting coffee growers and working with the beekeepers. I met Peace Corp volunteer Brad Roberts during my visit.

The regional agricultural center was building a modern extracting and honey processing center in Hojancha. A Development Bank loan was funding this and other projects in an effort to improve agriculture in the region. Ministry of Agriculture (MAG) beekeeping specialist Dauber Rodriquez was justifiably proud of this facility and the progress the beekeepers had made. Beekeepers in the region were seeking to increase their production by eliminating less productive and the more defensive colonies. Dauber identified the need to locally rear queens and improve requeening as the highest priority for continuing improvement to help insure enough honey to operate the new facility at an efficient level.

I was unable to revisit the cooperative in 1994 but did see honey with their label in the regional capital of Hojancha, when our tour visited the town. Talking with one of the beekeepers I learned that unfortunately honey production had not advanced and the queen-rearing project still had not started so the honey extraction facility was not operating. This region is the lowlands where Africanized bees are most defensive.

Paco Vasquez, Master Beekeeper

The 1994 beekeepers tour visited Francisco (Paco) Vasquez in the Central Valley town of Santa Ana. He was once the largest beekeeper in Costa Rica with over 2500 colonies; he now operates 700. I had visited him in 1984 and again in 1989. He nearly quit keeping bees in the mid-1980s after Africanization.

Paco Vasquez is one of those unique beekeepers. He is a soft-spoken, hard-working, excellent beekeeper. He moves smoothly and gently and it is a pleasure to observe him inspect a colony. He quickly assesses the condition of a colony with little wasted motion. He is one of the real artists among the beekeepers I have had the pleasure to meet and join in hive examinations.

In my three visits spanning 10 years, I have visited the same apiary of Paco. What is unique about this apiary is that it is located within 150 m (500 ft) of homes occupied by members of the Vasquez family. The number of colonies has been reduced from the 150 I found in 1984 but Paco still has 40 to 50 hives at the same location. Paco and members of the 1994 tour inspected his hives without gloves and even without veils at one point. (Figure 8–3)

Paco says he has had to change to survive in beekeeping. His honey production is considerably lower—about ¼ of what it formerly was. He has, however, developed a big demand for pollen which compensates for less honey. He sells all his honey and pollen at agricultural fairs and events in Santa Ana, a popular tourist and ceramic center in the Central Valley.

Figure 8–3. Examining Paco Vasquez's bees without gloves or beesuits.

Paco raises his own queens (using Italian queens from San Andreas Island queen breeder Ralph Newball as breeding stock) and requeens his colonies at least once, and sometimes more, each year. He marks queens and closely observes his colonies. If he finds evidence of defensiveness he requeens, particularly if the queen does not have one of his marks. He finds 700 colonies about the maximum he can handle with the necessary higher level of attention he feels important with Africanized bees.

F_1 Stock in Costa Rica

In Costa Rica, the production of F_1 queens and regular requeening has resulted in a strong and healthy beekeeping industry following Africanization. The beekeepers have changed by learning the importance of requeening. They buy queens which helps make the queen rearing business profitable for others. Everywhere the message is the same—change and adapt or get out of the business.

Costa Rican queen breeders repeatedly pointed out the weaker condition of the European-race colonies they maintain for their source of grafting larvae. They consider European bees poor producers compared to Africanized bees but believe the F_1 bees are more competitive. After looking at a few of the colonies, the U.S. beekeepers on the 1994 tour agreed that European-race colonies had less brood, fewer honey stores and generally were weaker.

Is the F_1 bee really a different bee? Jerry Loper of the USDA Bee Lab in Tucson measured the frequency of MDH (malate dehydrogenase) and HK (hexokinase) alleles in feral colonies, progeny from queen mother colonies, drone-source colonies and from daughter queens in the apiaries of two Costa Rican beekeepers who were rearing and selling queens. MDH represents a baseline and the frequency of HK has been demonstrated to differ in European and Africanized bees enabling scientists to use it as an indicator of bee race.

Loper's analysis revealed that allele frequencies of queen mothers in mating apiaries were European or European-African while drone-source colonies were more African in their allozyme frequency. The resulting F_1 colonies yield "a high degree of European allozyme frequencies" and colonies that are European "in behavior and appearance."

In a more recent study, Lobo compared wing morphology, isozymic variation and mtDNA in feral bee colonies from two lowland locations (Guanacaste and Golfito) to bees in the Central Valley. Although he found Costa Rican bees to be "predominately African," both isozymic and mtDNA data showed more European influence in bees sampled in the Central Valley compared to those collected in the Pacific lowlands.

These studies confirm what all the beekeepers and queen breeders of Costa Rica know—there is a higher degree of hybridization in the Central Valley region of Costa Rica and bees are more manageable at the higher elevation. Differences may be due to the importation and use of European queens or simply the different environment. The bottom line—more manageable and less defensive bees in the higher elevations compared to the bees at lower elevations.

Pollination

One of the most dramatic changes in beekeeping in Costa Rica has been the development of pollination rental opportunities. Large international companies, using modern irrigation systems and the latest technology, are growing melons in the Costa Rican lowlands for export to the U.S., Israel and Europe. Their markets have narrow windows for acceptable prices. Quality is foremost in importance, though quantity is also a factor in order to command favorable shipping rates and space.

In Costa Rica, beekeepers like the Perez brothers and Max Gonzalez rent bees to melon growers for $100 per colony. Rental consists of four moves of approximately two weeks duration each. Colonies are on pollination sites for two weeks and then moved away to forage for a two-week period before movement back to another portion of the field for another two weeks of pollination duty. With the short life cycle of Africanized bees, a two-week period away from the rental site is sufficient for complete turnover in the foraging population and allows colony populations time to recover from pesticide exposure and the poor melon-foraging conditions.

Most bee colonies then repeat the cycle for another month after which the colonies are moved to the Central Valley for the beginning of the coffee bloom. Virtually all colonies are moved by hand as most Costa Rican beekeepers lack mechanical lifting or moving devices.

The melon producers require colonies with decent populations in two boxes. Bee colonies are moved at night. Within-field distribution is needed for uniform pollination. Since wind is a constant feature, rows of melons are interspersed with a tall, protective vegetation barrier, usually King grass, and colonies are placed on the protected side of the barrier. Four colonies form a group spaced at approximately 300 m (1000 ft) intervals. (Figure 8–4)

Figure 8–4. Africanized bee colonies on melon pollination.

USDA. researchers Bob Danka and José Villa, with assistance of Norm Gary, investigated foraging distances of European and Africanized bees in this melon growing area in January and February, 1992. The study compared European-race bees, the F_1 hybrid (bees produced by Max Gonzalez) and Africanized bees selected using the same behavioral characteristics Marla Spivak used in her study of Africanization of Costa Rican bees. The European-race colonies were imported from the U.S. for this study.

These side-by-side comparisons revealed that Africanized bees forage a shorter distance from their hive than either European-race or hybrid bees. (Figure 8–5) When pollen foraging was analyzed separately, the differences were even more pronounced. The hybrid (F_1) bees supplied by Max were intermediate but more closely resembled Africanized than European bees.

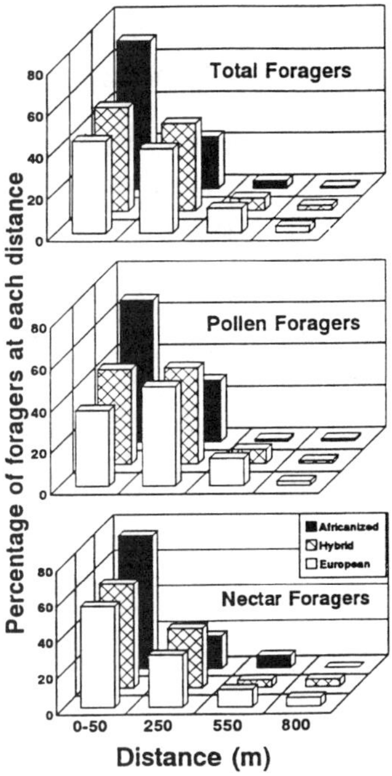

Figure 8–5. Percentages of bees (for 3 bee types)foraging at four distances in cantaloupe pollination. (Source: 1993 BeeScience *3:16)*

These findings demonstrate the importance of distribution of Africanized bees within a crop for pollination purposes. Africanized bees need to be closer to the flowers. This means more work for the beekeeper and an increased probability of field workers being stung. Thus the growers demand strong but gentle colonies and in-field distribution. As Miguel Perez and Max Gonzalez both emphasized, pollination rentals are not an easy way to make money, though the rental fees are high.

The use of honey bees for melon pollination in Costa Rica and the opportunity to obtain decent prices for rental colonies is not widespread in other tropical areas where Africanized bees now occur. However, as pointed out in recent reviews by David Roubik and myself, the Africanized bees now represent a new opportunity for expansion of crops dependent upon honey bees for pollination. Beekeepers in Costa Rica (melons) and Bolivia (sunflowers) have found pollination rental income an opportunity to continue with bees.

✧✧✧✧✧✧✧✧✧

While many Costa Rican beekeepers have quit and the number of hives is down, beekeeping is in a better condition and has recovered to a greater degree from Africanization than the beekeepers in Panama and other countries. One reason for the difference is the level of resources committed to assisting beekeepers. Costa Rica has a functioning system of regional beekeeping specialists employed by the Ministry of Agriculture that help advise beekeepers on bee diseases and bee management. The two major universities, University of Costa Rica and National University, have bee specialists who offer courses and conduct ongoing studies on honey and stingless bees. The Africanized bee requires change and adaptation—good educational resources help foster beekeeper survival!

Change, such as regular requeening and adaptation, is evident in the once-again expanding bee industry in Costa Rica. Pollination rentals and sale of pollen and queens are now providing income to beekeepers. Among Costa Rican beekeepers one senses an optimism that selected F_1 Africanized bees are a different bee. The beekeepers seek to husband selected F_1 hybrid bee considering them more manageable while still competitive with the feral Africanized bee population that now completely inhabits their territory. Apiaries are still maintained close to occupied homes and animals in the higher elevations.

9

Beekeeping in Bolivia

Bolivia is a landlocked country in the middle of South America. It is bordered by Brazil on the east and north, Peru and Chile on the west and Paraguay and Argentina on the south. At 1,100,000 km^2 (424,600 mi^2) it is slightly larger than Ontario Province and double the size of the state of Texas, minus the panhandle. (Figure 9–1)

Bolivia has about 6.5 million inhabitants. It can be divided politically and geographically into three regions (Departments), each with a large populous city. La Paz is on the altiplano—at over 3600 m (11,800 ft) it is the world's highest capital. Cochabamba, in the middle, is in a large Andes high-mountain valley (2558 m – 8400 ft). Santa Cruz (420 m – 1380 ft) is in the lowland Amazonian tropics. Bolivia, although considered an Andean country like Peru and Ecuador, is ⅔ tropical Amazonian lowlands. Bolivia, after Haiti, is the poorest country in the Americas.

Figure 9–1. Bolivia. (Source: The World Factbook)

Africanized Bee Arrival

The earliest record of western honey bees in Bolivia was importation of Italian bees into Santa Cruz in 1828 from Brazil.The Africanized bee arrived in Bolivia in 1967. Five years after Africanized bee arrival, the small but well-organized honey bee industry in tropical Santa Cruz reported a honey production decline of 40% (from 200,000 kilos – 441,000 lb annually). Colony numbers dropped to 1000, ⅕ the number of colonies prior to Africanization.

Noel Kempff Mercado was a student at the University of São Paulo in Ribeirão Preto, Brazil, when Professor Kerr carried the African bee to Brazil. He became a distinguished Professor of Botany at the regional Gabriel René Moreno University. and was a beekeeper and later President of the Beekeepers Association of Bolivia. He was one of only a very few tropical Bolivian beekeepers who continued to keep bees after Africanization.

Ten years prior to the bees' arrival, at the time beekeepers in Bolivia were hearing about the new bee from Brazil, Professor Kempff Mercado warned of the danger of importation of "Africanas" queens from Brazil because of their "extreme aggressiveness" and "tendency to swarm and migrate" (abscond). He also believed them good honey producers and "prolific" in their brood-rearing tendency, but their defensiveness, swarming and absconding tendencies, their susceptibility to European foulbrood and their failure to accept foundation (as also noted in Brazil) made them undesirable.

According to Kempff Mercado,

> "Wherever *adansonii [scutellata]* arrives everything falls into confusion and most people give up beekeeping.
>
> "These wild colonies can and must be destroyed by any means, but usually it is difficult to find them. The colonies must be destroyed by burning them, or by starving them by blocking hive entrances; it is not advisable to poison them, because this could also kill the gentle bees in the apiary, which would be attracted to the poisoned honey."

Outside of Bolivia the beekeeping world knows little of Professor Kempff Mercado or his writings about Africanized bees. He was tragically killed, after only 15 years as Botany Professor, while leading a botanical expedition in the jungle of northeast Bolivia, looking for plants, including plants that bees visit—one of his specialties. His expedition members were killed because they inadvertently got too close to a jungle cocaine processing plant. His death led to a great public outcry against cocaine and drugs, which were then widely promoted by a corrupt military government. Today a large national park is named in his memory, ironically in the very region where he was investigating honey plants. You can see a memorial to him in one of the many parks in the city of Santa Cruz.

A Tropical Beekeeper—Francisco Kempff

I had the opportunity to visit and talk with Francisco Kempff, the son of Noel, also a beekeeper. He remembers his father's bees near the house when he was a youngster. His present apiary locations are all in the country, away from people and animals. He says his beekeeping is completely different from that of his father, from necessity. He stresses that beekeeping with Africanized bees is not "strange" but anyone who keeps Africanized bees must treat the bees with "utmost respect."

Francisco is not a full-time beekeeper, though he has kept as many as 250 colonies at one time. He currently maintains about 50 honey bee and 10 stingless bee colonies. He arranges with rural campesinos for care of his colonies. He rents a house in the country and keeps his bees near the house, sometimes within the adobe walls of the patio. The campesino watches the colonies in return for the first kilogram (2.2 lb) of honey harvested at the apiary location.

Francisco's system is necessarily simple. The campesino caretakers feed the colonies dry sugar during the rainy season to reduce absconding. To replace his losses, Francisco captures reproductive swarms at the beginning of the dry season in bait hives using a piece of comb as bait. He selects colonies that are productive and less defensive and raises new queens from such colonies by dequeening so the bees raise emergency queen cells. Some queen cells are left (to requeen the colony) and others removed to requeen other non-productive or overly defensive colonies.

He believes part of the secret of avoiding defensive behaviors in Africanized bees is in the handling of the bees. Thus he makes his own equipment, to insure that frames fit properly which enables him to inspect colonies with a minimum of jarring or bumping of the hive. Since Africanized bees use lots of propolis, poorly fitting equipment is a detriment. He prefers certain native woods, drying the cut wood well before making the boxes and frames.

Francisco performs colony inspections with gloves, coveralls and sturdy veil. He uses the big smoker from Brazil; the campesino helper continually smokes the bees during inspections. Francisco prefers to work in the morning hours, as the afternoons can be very hot, humid and windy; his father, 20 years earlier, recommended working Africanized bees very late in the afternoon.

Francisco's beekeeping operation is currently suffering from an experiment that nearly wiped him out. He purchased $12 Italian queens from Canada along with $2 Italian queens from Chile and introduced them to his colonies. He described his trial as "a complete bust." He found that colonies with Italian queens built up slowly, used the honey flow to rear brood, and then stored little surplus. "The Italian bee is simply not competitive with feral colonies of Africanized bees" according to Francisco.

With Africanized bees, Francisco produces an average of 30–50 kg/colony (66–110 lb/col). One thing he has learned in Africanized bee management is to remove honey-filled frames, extract and then return the frames the same day. His honey is light and comes mainly from clovers and alfalfa grown by Mennonite dairy farmer.s

Isaac Nuñez

Isaac Nuñez (Figure 9–2) is the dean of beekeepers in Santa Cruz, one of about 100 currently active beekeepers maintaining about 3800 colonies in the Santa Cruz Department. He has been a beekeeper for over 35 years and attributes his energy and youthful appearance (he is 76 but barely looks 50) to his work with bees and consumption of bee products. Isaac currently manages 250 Africanized colonies; he maintained over 600 Italian bee colonies before Africanization in the late 1960s.

Figure 9–2. Isaac Nuñez with observation hive and extractor at the National Honey Festival.

Isaac is migratory within about 100 kilometers (60 miles) of his home and extracting facility in Santa Cruz. His seasonal management is very similar to a calendar recommended by Professor Kempff Mercado shortly before he was killed. During the rainy season, Isaac keeps his bees in an area where a native tree (parjobobo) provides a dark honey, when the rain relents enough for the bees to exit their hive.

In March, Isaac transports his colonies to sunflower locations along the Rio Grande River. Sunflower, introduced about 10 years ago, is a new crop for Bolivia. Isaac leaves his bees in fields of sunflower for 2–3 months (March–May), though individual fields only bloom for about 3 weeks. Isaac rents his colonies for US$25 per colony; some growers obtain bees as cheaply as US$15.

The honey of sunflower crystallizes rapidly and must be promptly extracted. Isaac additionally needs to reduce colony size to move his colonies from the sunflower fields to locations around Santa Cruz, where the bees will forage a succession of flowering plants during the dry season. He moves only at night and not long distances. He says it is possible to move Africanized bees but it is harder. He plans his moves carefully.

Varroa has been present since the 1970s but Isaac does not treat his colonies. European foulbrood (EFB) is present but it is not very common. Ants and termites are more serious and persistant problems. Isaac is trying to determine if the bottom portion of plastic soda bottles, inverted over the four wooden stakes of an individual hive stand, will control both insects.

Isaac has observed a change in the honey flow since the bee population in tropical Bolivia changed from European to Africanized. Now the honey flow comes in April to June (the winter), whereas before it was in August to November (the spring). Isaac finds he must reduce the amount of swarming in his colonies to insure a honey surplus. He calls swarming "one of the most negative factors" of Africanized bees.

Isaac does not remove bee swarms except as a public service. He says they are hard to relocate and often do not remain in equipment. Swarms early in the season (May) are easier to capture and husband; those near the end of the season are often not useful. We examined a November swarm (probably an abscond) he had just removed at the airport. Since the small colony was rearing brood within two days, he predicted it might stay.

Isaac has tried different bee stock to improve his beekeeping return over the years. He has tried European queens on more than one occasion but says "such colonies do not perform well." For a while, following Africanization and a visit of Dr. Allyn Stearman, a Florida hobbyist beekeeper, the Central Florida Beekeepers Association "adopted" the Santa Cruz Beekeepers and sent free Italian queens to Isaac and other Santa Cruz area beekeepers in Bolivia. He has also tried European-race queens from Chile and Argentina, has imported Africanized queens from Brazil and even reared his own. He is satisfied with the current queen stock (the so-called improved Africanized bee—see Chapter 10) imported from Brazil.

Isaac is typical of surviving Santa Cruz beekeepers in that he has learned to use his bees to advantage. He moves his colonies during the year, and he rents his colonies for sunflower pollination. He also reports a 50 kilo (110 lb) harvest, double the average. In each country or region there is an Isaac Nuñez—a beekeeper who has adjusted and learned to profitably survive keeping Africanized bees!

National Honey Festival

I visited Santa Cruz during the Second National Honey Festival. The festival had a distinct flavor of Santa Cruz but the objectives of focusing attention on bee products and attracting the general public to see honey bees in their most favorable light was the same as for any festival elsewhere. In Bolivia, these objectives are especially important since the only honey bee beekeepers have to manage is the Africanized bee and widespread honey adulteration is suspected by consumers.

The Bolivian honey festival was held in a grove of trees on the farm of beekeeper Ovidio Mercado in the community of Gavita, an hour outside the regional capital of Santa Cruz. A local association of about 15 beekeepers were co-hosts along with the larger regional bee association of Santa Cruz. The festival itself was organized by Isaac Nuñez.

While participating in the Festival, I visited Ovidio's apiary but we did not open colonies as this could have resulted in festival participants receiving stings. Ovidio had 65 colonies of honey bees in the jungle around his house, nearly all in standard movable frame hives. Ovidio said he averages about 40 kilos (88 lbs) per colony. He also had 40 boxes of the native stingless bees from which he harvests up to a kilo (2.2 lbs) of honey.

Ovidio showed me his honey house. (Figure 9–3) The structure, typical for the tropical region, consisted of bamboo slats covered with adobe for walls with a palm-frond roof. Ovidio had added fly screening at holes and over the window and door to keep bees out of the extracting area. He harvests 3 to 4 times per year, typical of tropical areas. He extracts at night because of the potential robbing behavior of the Africanized bee. Although larger than average, his apiary and management were typical of tropical beekeeping in Bolivia.

Figure 9–3. Ovidio's honey house.

Tropical Beekeeping in the Yungas and Chapare

The two other most heavily populated Departments (states) in Bolivia, La Paz and Cochabamba, include tropical regions within their sector. Both areas have become household names within and outside Bolivia because they happen to have the right climate, soil, and heavy rainfall conditions that are ideal for the growing of coca (*Eritoxylon coca*). This crop has long been used by the natives of the high Andes altiplano to combat the severe climate and hold off hunger but it also forms the base for cocaine that has become an all-too-popular world-wide drug.

The tropical areas, the Chapare of Cochabamba Department and the Yungas of La Paz, are located at the base of the front range of the Andes Mountains. Since beekeeping is not very profitable within the Andes, except in some of the larger valleys such as Cochabamba, these areas also represent some of the best areas for honey production to serve the adjacent large-city markets. Beekeeping might also represent a viable alternative for the campesinos to the growing of coca.

The Yungas is north of La Paz and is only reached after a harrowing 4- to 6-hour drive on a twisting, turning highway that plunges from 4000 m (13,125 ft) of the altiplano to under 1000 m (3000 ft) of the tropical Amazonian region. (It is not a trip for the faint-hearted, especially in native transportation.) A trip to the Chapare is easier but with a large U.S. military presence and active coca eradication programs it is often a volatile political area. With abundant flowering vegetation both are excellent beekeeping areas.

The internal market has been changing in Bolivia with larger national/multinational companies taking over small firms as in the rest of the world. It is difficult for rural, campesino-oriented organizations to successfully market their products. With United Nations Food and Agricultural Organization (FAO) assistance, Oscar Arispe, through his private enterprise Apisbol, has started women's beekeeping groups in several Yungas communities. Apisbol financed the entire start-up expenses of the participants—all the beekeepers paid off their loans with honey harvests within the first four years.

Another group, Agroyungas conducts a large rural development program that funds projectss for alternatives to coca culture. Funds (estimated at nearly US$5 million annually) come from the EU (European Union). Agroyungas offers credit to individuals to begin 2 to 5 colonies with honey harvests dedicated to loan repayment. In the Chapare FAO funds similar programs while USAID (U.S. Agency for International Development) has a large alternatives program with Oscar (see page 121).

In non-coca growing regions the campesinos do not have access to funding programs like Agroyungas or USAID money. Cecilio Cardenas of a beekeeper cooperative, Agromiel, was successful in obtaining US$6500 to start 100 colonies with 10 associates from a NGO (non-governmental organization) called FADES (Fundacion para Alternatives de Desarrollo) in the Yungas. The group was able to keep costs low by locally capturing swarms and making equipment within the communities.

Initial plans to manage the bee colonies cooperatively did not work well so the group evolved into a commercial association. The group prospered because the area proved to be a good one for bees (average yields are 40 kg/yr – 90 lb), and markets were established in La Paz. Today, 30 members manage over 660 colonies.

Beekeeping in the Bolivian High Andes

Cochabamba is a city of nearly one million, located in a high mountain valley (2558 m – 8400 ft elevation) behind high peaks of the eastern front range of the Andes Mountains. Cochabamba did not have the western honey bee until 1858 when they were imported from Chile. We do not know precisely when Africanized bees arrived in high mountain valleys such as Cochabamba but it was after 1970.

Today there are 4000 to 5000 colonies in Cochabamba and the surrounding area, an estimated 60% in non-movable frame (rustic) hives. Most of the beekeepers have small numbers of colonies though there are some larger migratory beekeepers which move bees from the high valley to tropical areas of the Chapare and back again.

Typical of the non-migratory beekeepers is my brother-in-law Alberto (Bedo) Narvaez. He has 5 colonies at his home in Tiquipaya, tucked up against the northern mountains of the main Cochabamba valley. Bedo obtained the bees from his grandfather Tames who was a beekeeper as long as anyone could remember. Bedo transferred the bees from the city of Cochabamba to his home at the base of the mountains to escape urbanization and to take advantage of the irrigated flower-growing fields in Tiquipaya. (Figure 9–4)

Figure 9–4. Bedo's apiary.

Bedo's colonies will store surplus honey in February, March and April and Bedo will harvest honey 2 to 3 times. Before this season, Bedo harvested 20 to 40 kilos (44–90 lb) per hive, selling the honey at the family-owned neighborhood store.

Beekeepers like Bedo in high valleys of the Andes Mountains, such as Cochabamba, have a long dry season to contend with when there is little forage for the bees. Beekeepers in areas where there is little irrigation secure yields of only 10 to 20 kilos (20–44 lbs) per colony.

Migratory beekeepers such as Julio Coca move an estimated 1500 bee colonies from the valley to locations on the tropical side of the front range of the Andes (the Chapare) to secure honey crops. Julio currently is expanding from 120 colonies. He averages about 50 kilos (110 lb) of honey from tropical locations in the Chapare but only manages about 15 to 20 kilos (33–44 lbs) in the valleys around Cochabamba.

Julio has been a beekeeper for about 10 years. Before beekeeping he was in the chicken-rearing business. He started with ten rustic colonies and found he preferred beekeeping to raising chickens so he transferred his rustic hives to modern Langstroth boxes, which he constructs himself, and began to expand.

In the tropical Chapare region, Julio's bees store honey from abundant native vegetation June until November before arrival of the heavy rains. When the season finishes and harvest is complete, Julio moves the colonies to the high valleys of the Andes Mountains like Cochabamba to take advantage of the nectar/pollen flow brought on by the same November–December rains. The last harvest will be in March in the valley region before colonies are moved back to the tropics.

Julio produces more honey by moving a short distance since seasonal nectar/pollen availability in the tropical areas follows the cycle of the Andes Mountain plateau. Within the Andes there are hundreds of valleys and each is different from the other. Generally, during the rainy season of the tropics, bees store little surplus but when the rains cross over the mountain peaks they bring needed water to the dry mountain valleys triggering plant growth, flowering and a nectar flow. Beekeepers like Julio have learned to seasonally manage colonies with such cycles.

Julio also produces pollen, propolis and royal jelly with his colonies so he is really managing several cycles all at the same time. Julio sells his entire production directly to consumers. He charges 15 Bs (about US$2.00/lb) for the best quality, lighter honey and slightly less for the darker honey. Pollen and products incorporating royal jelly or propolis, made especially for him by a private lab, are more profitable. He would like to produce more pollen since demand is high but the commercial pollen traps are not effective for the smaller-bodied African bee.

Julio says the local market is generally very good. One recent study showed honey demand far exceeded supply. Julio promotes his honey as ecological since his nectar sources are all from natural vegetation.

By moving his colonies, he obtains greater production from the same colonies, leading to a year-round supply.

In the high elevation valleys his bees are moderate in temperament; the same bees are considerably more defensive at tropical locations. In the tropics the bees must be heavily smoked and worked with care to avoid bumping and spreading of alarm odor to other colonies. By seeking more isolated sites and working carefully, Julio has not had too many accidents, though he did have one instance where his bees killed 12 cows after he finished manipulations. (He "bought" the dead animals.)

One thing Julio has discovered, common for beekeepers around the world, is that success quickly leads to imitation. His locations in the tropical region were highly productive two years ago but this year have been less so as nearly a dozen other beekeepers now also move their colonies to the same general area. Local beekeepers, some with assistance from U.S. Peace Corps volunteers, at times create a sharp conflict between migratory and locally resident beekeepers. Too frequently the migratory beekeepers are forced to move on after demonstrating the potential of a new area.

Oscar Arispe

Mention bees or beekeeping in Cochabamba, and immediately people ask if you know Oscar Arispe. Oscar is the owner and chief salesman for his business Apisbol. He is a beekeeper, though he finds little time to manage his bees. His main activity is selling and promoting beekeeping supplies, most made in Cochabamba itself, and an extensive line of health and beauty products made from bee products. (Figure 9–5) He calls the bee hive "the local medicine cabinet." He never tires of promoting the benefits of his products.

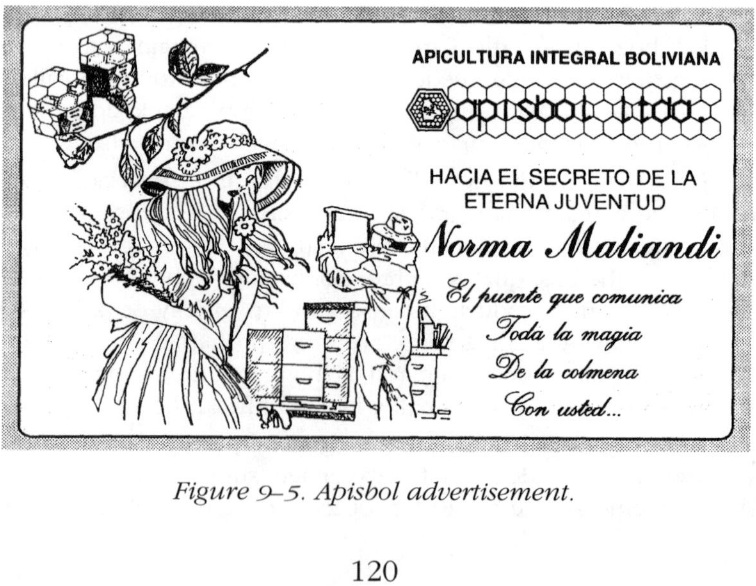

Figure 9–5. Apisbol advertisement.

Oscar has been very active in promoting campesino beekeeping development projects. He developed a beekeeping program in the tropical Yungas region of Bolivia that has yielded more than two dozen large beekeepers. These individuals all paid off their loan within four years and their small honey businesses are growing. Now Oscar would like to promote beekeeping as an alternative to coca culture and do the same thing in the Chapare as he has in the Yungas.

I joined Oscar, his son Daniel, and Oscar Sanchez of USAID (U.S. Agency for International Development) in visits to over 10 communities in the Chapare region in early October 1995, to discuss a new initiative promoting the concept of local campesinos becoming large-scale beekeepers. (Figure 9–6) USAID is willing to help finance this program.

Figure 9–6. Oscar Sanchez (USAID) and Oscar Arispe (right) meeting with beekeepers in Chapare.

Oscar's role is to provide beekeeping supplies, including the bees themselves, and conduct the beekeeping training. His method of teaching is straightforward, entertaining and informative. He is like a minister constantly preaching the "gospel of the bees." When visiting beekeepers in apiaries, as in Ivirgasama where we spent time in the apiary of Pascual Tonconi, Oscar seeks to motivate and encourage, while never hesitating to recommend changes and suggestions to the beekeeper.

In Mariposa Oscar and I visited the apiary of Ramón Momoni, President of Apitropolis, the largest organization of beekeepers in Chapare. The association consists of 40 persons who manage about 60 colonies for honey, pollen and propolis, which they market through Apisbol.

Ramón had one of his colonies in an unusual location—it was in the kitchen ground floor of his house. In this tropical region, many homes are two stories. The family sleeps on the second story where it is a bit cooler and away from animals (domestic as well as wild). Ramón had moved one of his colonies from the adjacent jungle to the ground floor kitchen and when I asked why he said because of problems with army ants. In his home he could "better look after the colony" and halt continued ant attacks.

The colony we looked at was strong, with good flight, and was filling frames Ramon had removed and harvested two weeks earlier. A second colony in the bush was extremely defensive. It too had abundant honey—in fact both colonies needed more room. The two colonies were swarms or transfers from tree trunks. Although Ramón was a good, knowledgeable beekeeper, he lacked proper equipment so he was more a robber of his bees than a beekeeper.

The opportunity to teach beekeeping to campesinos represents a return to the Chapare for Oscar. He was the first person to try large-scale beekeeping in the Chapare region with 250 colonies headed by gentle Italian queens from Argentina as the area opened up with a new highway for coca exploitation by the Bolivian military in the 1970s. His beekeeping venture fell victim to inappropriate timing. Civil unrest, an oppressive military dictatorship and poor economic conditions within Bolivia made frequent visits to the colonies difficult.

The bees became Africanized. Oscar had to abandon his bees. This time he hopes fortune will shine on his initiative!

Mizque

The Mizque valley is another picturesque high (1850 m – 6100 ft) valley of the Andes Mountains. It is south of Cochabamba on the opposite side of the 1.5 million acre Carrasco National Park from the Chapare. The valley is reached via a long, bumpy dirt road from Cochabamba. CEDEAGRO (Centro de Desarrollo Agropecuarias), an NGO (non-governmental organization) operates a beekeeping program that involves over 200 families in 30 communities in the valley, managing over 500 colonies, the majority in movable frame equipment. One of the reasons international organizations support CEDEAGRO is their demonstrated effectiveness in reaching the rural campesino family with economically viable programs such as beekeeping, so as to reduce out-migration of the rural population into cities like Cochabamba or the coca-growing Chapare region.

The Mizque beekeeping project began in the mid-1980s when the EU (European Union) and Elisa Foundation financed a US$22,525 investment. The families purchased bee equipment on credit and stocked it with locally available bee swarms and transfers of feral colonies from trees and rock caves. Mizque beekeepers produced 879 kg (1900 lb) of honey in 1991 (11 kg/col – 24 lb average). This increased to 3400 kg (7500 lb) from

172 colonies (20 kg/col avg. – 44 lb) in 1992. Production is now over 4 tons of honey plus pollen, propolis, and beeswax. Very impressively, US$14,874 of the original investment was repaid and over 40% of the families cancelled their initial debt in the program the first four years.

For these 200 beekeeping families, the Africanized bee has proven to be beneficial. With Africanized bee colonization, swarms are now plentiful, allowing Mizque valley families to inexpensively stock their bee equipment paid for with harvested product on a 50:50 basis—50% to retire the credit advance of equipment and 50% for their use. About 20% of the colonies are rustic but some families maintain modern movable frame hives alongside rustic hives; 10% of the beekeepers have expanded to more than five colonies.

Figure 9-7. Valerio Cayola bee technician of CEDEAGRO in Mizque apiary.

Part of the reason for the CEDEAGRO program success is the service provided to the individual beekeepers by technician Valerio Cayola. He extracts and sells the honey and assists them with hive management, working with the beekeepers on beekeeping skills. Valerio is an excellent beekeeper and a good, patient teacher with beekeepers. He carefully explains what is happening and helps the beekeepers move frames, requeen, add supers and perform necessary management.

Valerio often has to walk several kilometers to reach beekeepers. For some apiaries, he carries plastic boxes to the apiary, transfers full frames with honey to them and then walks to a location where there is water and electricity to extract the frames. Empty frames are promptly returned since strong colonies will soon need the honey storage space and, with Africanized bees, the colonies are often very defensive the following day. Beekeepers have learned that it is better to promptly replace frames.

The beekeepers help extract the honey to cut labor expenses. When Valerio extracts, he carefully weighs the honey and wax. For those who still have not paid the original investment, 50% of the value is returned directly to the beekeeper (or the honey itself if the beekeeper wishes) and the other 50% is credited toward their outstanding loan balance.

The Africanized bee, like all honey bees, is dependent upon local conditions for honey production. Ten kilometers (6 miles) south of the town square, the colonies were strong with large brood patterns and some had enough honey to harvest. In Buena Vista at the north end of the valley in the apiary of Florentino Claroz, we found a very different situation. His apiary of 15 colonies in a small grove of eucalyptus trees was in a much drier area and none of the colonies had expanded out of the first hive body. All lacked surplus stores but had abundant brood, a characteristic of the Africanized bee.

The Mizque valley project has tried Italian queens. The queens were a gift from CESO (Canadian Executive Services Organization—a volunteer organization similar to ACDI/VOCA in the U.S.). The Italian bee colonies built up more slowly than Africanized bees and used the nectar flow to build, whereas Africanized bees expand rapidly with pollen availability and are strong enough to use the flow to store surplus.

When I inspected colonies in the valley, it was possible to clearly see which colonies were the descendants of the original Italian queens. European colonies all had smaller populations, less brood, and little honey stores compared to colonies in the same apiary that were Africanized. European colonies also displayed fewer behaviors of Africanized bees—they were less nervous, more gentle, used less propolis, responded to smoke more readily and did not abandon their hive when opened.

Criolla versus Africanized Bee

Julio Coca, Oscar Arispe, the Mizque Valley beekeepers and other beekeepers in high-elevation Andean valleys like Cochabamba and Mizque believe their bees to be a mixture of bees brought by the early Spanish settlers and the Italian bee. They term them "criolla" bees. This belief stems from the less defensive nature of their bee colonies compared to the Africanized bees present in the lowland tropics.

To determine if the high-elevation valleys of Bolivia had a different bee from the tropical lowlands, I collected bee samples in two transects from low to high elevation. Colonies were examined and given a ranking based on their behavior as European, intermediate or likely Africanized. Adult bee samples were collected and later analyzed by Steve Sheppard and Anita Collins at USDA Beltsville, MD, for DNA and morphometric determinations.

Two colonies, descendants of European-race queens, a colony at CEDEAGRO in Mizque, and another colony from a queen from Denmark, in the apiary of Julio Coca, were European in DNA and morphometric measurements; the Julio Coca colony was Africanized in the field judgement (highly defensive) while the other was rated as very gentle. All the remaining colonies were African genetic material (in DNA and morphometric measures). Only one of these colonies, at mid to higher elevations was judged Africanized in field analysis—most were intermediate, with two rated as European. All colonies were considered totally Africanized in more tropical areas.

Thus the colonies captured in the wild, along with those managed by beekeepers, have African bee DNA unless they had been requeened within the past three years. They were not a mixture of European-race genetic material—not a criolla bee. But although the bees of Bolivia are Africanized, like the rest of the Americas, higher-elevation colonies do not display Africanized bee behavior comparable to the same bees found in adjacent tropical areas of Bolivia. This difference is also reflected in their still being kept close to homes, near other animals, and manageable with fewer changes on the part of the beekeeper. (Figure 9–8) It is readily apparent that higher-elevation bees are more manageable!

Figure 9–8. Bolivian highlands apiary (Cochabamba) still located near home post-Africanization.

A few years ago a large beekeeper, Berhard Haas of German descent, compared different bee stocks under local conditions in his apiary near my brother-in-law's bees in Tiquipaya of the Cochabamba valley. He stocked his apiary with 10 hives with Italian queens, 10 with Africanized queens brought from the tropics and 10 with queens from the bees he was presently maintaining in his apiaries. He inspected and managed all of them in the same way, and kept track of their honey production. Within one year it was evident that the local bees were superior, storing the greatest amounts of honey in this test; Italian bees stored very little surplus. This trial confirmed what others have found—local bees do better.

Haas, Julio Coca, CEDEAGRO in Mizque, Francisco Kempff in Santa Cruz, and many other beekeepers in Bolivia have repeatedly demonstrated the futility of attempting to reintroduce Italian bees to areas with high populations of Africanized bees. Italian queen colonies do poorly and are difficult to manage for honey production. In repeated trials, honey production from colonies headed by selected or feral Africanized bee colonies always out-perform Italian bee colonies. European bees are found to be unsatisfactory!

National Beekeeping Congress

The first National Beekeeping Congress in Bolivia was held February 20–22, 1995, with 60 persons attending. The meeting was sponsored by ADAC (Asociacion de Apicultores de Cochabamba) and Food and Agricultural Organization (FAO) of the United Nations. Beekeeper and Cooperative President Luis Cordoba of Uruguay was a special guest. He spoke on several topics, including 1976–80 FAO funding that promoted strong cooperatives (recently strengthened by Interamerican Development Bank grants) to develop the beekeeping industry in Uruguay.

The Congress stimulated development of an ad-hoc committee to approach FAO to ask for beekeeping development assistance along the lines Uruguay received. The group is currently formulating a national proposal. To address another major issue of concern, the Congress called for formation of a National Institute of Apiculture (INA) that would help formulate a law in Bolivia to cover bee health/disease control, regulate apiary locations, study bee flora, address honey adulteration and work on certification of honey origin and bee ecological products.

Drs. Gonçalves, Stort and De Jong stated that the first national meeting organized for beekeepers in Brazil (150 attended the initial Congress in Florianópolis) was "of historical importance" for discussions of "new methods for management of Africanized bees." This Congress helped lead to "full development of Brazilian apiculture." Hopefully, the first Congress for Bolivia will likewise positively impact the future development of beekeeping to the advantage of Bolivian beekeepers.

✧✧✧✧✧✧✧✧✧

Beekeeping has not completely recovered from colonization of Africanized bees in tropical Bolivia. The tropical areas, larger than the size of Texas, have great potential when transportation and access become available, an important factor that has enabled beekeeping to expand into tropical areas in neighboring Brazil. Africanized bees become a ready resource (abundant swarms and feral colonies) with development.

There are fewer colonies today and virtually no hobbyists in tropical regions of Bolivia, but beekeeping is different in the Andes Mountain region of Bolivia compared to the tropics. Beekeepers must maintain Africanized honey bees, which they call a criolla bee, but bee colonies are more manageable and problems with stinging accidents are reduced. Most of the high-elevation valleys are dry and honey yields are low though pollen harvest can be substantial. Many apiary sites have not changed in the highlands.

Private bee supply businesses like Apisbol and NGOs like CEDEAGRO in Mizque are assisting rural campesinos in both tropical and high-elevation communities become beekeepers by providing training and equipment loans. Many loans are successfully repaid with harvested honey. Educational efforts and development of a national organization are beginning to supply the growing need for beekeeping information and national recognition and support for beekeepers.

10

Africanized Bees in Other Countries

As evident in the preceding chapters, Africanized bee impact has been different in Panama, Costa Rica and Bolivia. This chapter, briefly examines Africanized bee colonization impact in five additional countries of the Americas. The story is somewhat different in each country.

Belize

Belize, formerly British Honduras, is the smallest Central American country. It is 23,000 km^2 (8878 mi^2) in size (slightly larger than Massachusetts). It forms the southern flank of the Yucatan, bordering the Peten region of Guatemala on the west and north and a bit of Mexico on the east. It has the lowest population density (21 indiv./km^2) of any of the countries of Central America and still has strong economic ties to England.

Belize was a major center of Mayan culture before European colonists arrived in the Americas and Mayan ruins and many Mayan descendants are in the country today, especially in the rural southern Toledo district and inland from the coast. Belize has the largest barrier coral reef in the Americas and tourism is an important industry. In the drier north (about 150 cm [60 inches] of rain annually), sugar and rice are the major cultivated crops along with citrus. In the wetter south, citrus, banana and forestry are significant income generators.

Henry Mulzac, a former Peace Corps beekeeping volunteer, dates beekeeping in Belize only from the 1950s. Statistics indicate that the industry grew steadily during the 1960s and 1970s. This expansion was financed through U.S. Peace Corps, CARE, USAID and CIDA (Canadian International Development Aid). A Beekeeping Extension and Research Project (BERP) trained and equipped over 300 new beekeepers and provided a National Beekeeping Center in the capital of Belmopan. In 1980, there were around 500 beekeepers keeping 10,000 colonies.

The two northern provinces are the best areas for bees and where the majority of honey bee colonies are located. Beekeeping in Belize has traditionally been a private enterprise centered around a dozen cooperatives. Mennonites are the major beekeepers in the north, Mayan Indians in the south and interior. Nearly all the honey produced (80%) is exported to Great Britain through the Belize Honey Producers Federation of Cooperatives.

Belize began to survey for Africanized bees in 1987 as beekeepers reported changes in their bees and Africanization occurred in Mexico. The arrival of Africanized bees was confirmed in March 1987. By September of the same year, 343 Africanized bee swarms were found in pheromone-baited "piñata" traps. With short north-south distance (280 km – 174 mi) from one border to the other, Africanized bees spread quickly in Belize.

The estimated number of beekeepers dropped from 500 to 100 following Africanization. Honey production, once at 270 tons, fell below 36 tons in 1992. Production per colony remains high though hives now require better management to produce decent honey yields.

Mike Burgett of Oregon State University visited 58 apiaries containing approximately 620 colonies during six weeks in early 1990 as part of an ACDI/VOCA[1]-sponsored development project. He reported the "majority of apiaries examined were in a state of literal and figurative decay" with much "beekeeper discouragement." Field identifications revealed that "nearly all apiaries and most colonies in those apiaries showed positive signs of mild to severe Africanization."

Villanueva, a Mexican bee scientist, conducted an independent survey of beekeepers in the two northern districts of Orange Walk and Corozal the next year These two districts border Mexico and contained 75% of the colonies in Belize prior to Africanization. A total of 116 beekeepers owning a total of over 2500 hives were interviewed. About 50% of the beekeepers "abandoned their hives" as "little or no instruction on how to manage Africanized bees or equipment for personal protection was available in Belize." Honey production was down 70% in these two districts. (Figure 10–1)

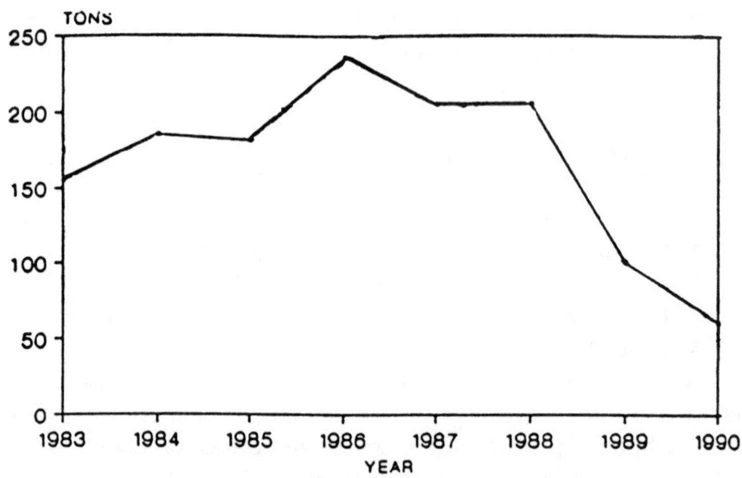

Figure 10–1. Honey production in Belize.
(From 1991 American Bee Journal *131:185)*

---

[1]ACDI/VOCA—Agricultural Cooperative Development International/Volunteers in Overseas Cooperative Assistance.

Based on colony visits and discussions with beekeepers, Dr. Burgett predicted that "beekeeping will remain on the decline in Belize...[if]...left unassisted." A few beekeepers had adapted to the Africanized bee by changing management systems and apiary locations. He optimistically stated that "in 10, to 15, to 20 years beekeeping will perhaps regain the economic stature it held prior to the entrance of AHBs."

Two of the recommendations from the visit, were:

(1) "immediate development of European queen sources,"
(2) "establishment of an indigenous queen-rearing program."

Eventually 5000 Italian queens were imported from Hawaii to requeen and start a queen-rearing program under the capable direction of Martin Cal of the Belize Honey Producers Federation. Unfortunately beekeepers reported poor success with queen introduction. Successfully requeened colonies did not respond well to the subtropical conditions of Belize.

Marijuana Defoliation and Honey Bees

The two surveys of Burgett and Villanueva were requested and partly supported by the National Belize Honey Producers Federation. Due to the drastic decline in honey production, the Federation suddenly was unable to supply their secure honey market in England. Some beekeepers, not acknowledging Africanization, believed the reason for the decline in honey production was the extensive aerial spraying of marijuana plantings.

The anti-marijuana spray effort, conducted by and paid for by the U.S. DEA (Drug Enforcement Agency), was highly unpopular within the country. The program, started in 1983, utilized aerially-applied herbicides to kill the marijuana fields. Tom Webster from Kentucky and Robert Danka of the USDA Bee Lab in Baton Rouge, LA, invited to Belize to evaluate beekeeping, stated that the marijuana spraying was not the major reason for the decline in honey production.

ACDI/VOCA sponsored a longer visit by Oregon State scientists Mike Burgett and Glenn Fisher. In a series of tests, bees were fed the herbicides used to spray the marijuana fields at concentrations well above those used in the field and bee colonies were moved to areas being sprayed (at some risk considering the controversy of the spray program). These scientists also were unable to find any negative effects of the herbicides in either lab or field tests.

Southern Belize

I had the pleasure to visit and work in the southern Toledo district of Belize among the communities of the Mayan Indians. The beekeeping in this region, a 12-hour dusty, bumpy bus ride from the capital, is more basic and less-developed than in the northern and western regions of the country where the industry resembles Yucatan beekeeping of Mexico.

Beekeeping began in the Toledo district largely through the efforts of U.S. Peace Corp volunteers. Prior to the 1970s, there was virtually no beekeeping in this area—a 1972 census listed 2 beekeepers with 64 colonies in the district. Beekeeping was a new activity for Mayan subsistence farmers.

I lived and worked in Toledo district with Peace Corp volunteer Tom Hyden. The Africanized bee population had arrived in southern Belize one or two years before Tom. Initially, Tom helped beekeepers move their apiary locations to more remote sites. His biggest challenge was "finding ways within the Mayan culture to solve problems." He believed "change has to come from within."

Figure 10–2. Peace Corp volunteer Tom Hyden with Mayan beekeeper.

With several languages continuously in use in this area of Belize, public meetings are not always an effective educational tool. Tom and I worked mostly on a one-to-one basis with the beekeepers. We would visit the beekeeper to gain his confidence and then would make arrangements for a return visit to inspect bees with the beekeeper. Most Mayan communities do not have telephone or electrical service so such visits are the only means of communication.

Like most Peace Corps volunteers, Tom wondered if he really made a difference. He related that "It's taken me two years to work these [Africanized] bees. I've made every mistake and been paid for them with stings, but I also work bees 3–4 times a week. These beekeepers only see their bees once or twice a month. The learning will take longer."

Virtually all beekeepers in Toledo district are members of the Southern Beekeepers Cooperative Society founded with Peace Corps assistance in 1980. The Southern Coop sold almost all the honey that members produced to the Belize Honey Producers Federation of Cooperatives, principally for international export. The Co-op record clearly shows the Africanization process in southern Belize. [Africanized bees arrived in late 1986–early 1987—by 1988 the honey handled by the association was less than half the previous year and only about one-fifth the level of the best year of 1984. In the last year of good figures (1990), production was a mere 818 kilos—1800 lb].

Table 10.1

Year	Production Of Honey (kilos)	Number of Colonies	Number of Beekeepers Selling
1980	30,000	1000	63
1981	20,840	1000	73
1982	27,000	1184	83
1983	23,270	1130	79
1984	32,280	1110	69
1985	19,430	1106	69
1986	16,100	1115	29[1]
1987	16,100	860	67
1988	5940	514	39
1989	2440	232	18
1990	818	200	10

[1]A number of beekeepers sold honey privately outside of the co-op according to Peace Corps volunteer Tom Hyden.

In 1986, as beekeepers were struggling with a lack of working capital, the co-op was fortified with a grant from Canada and another from the International Development Index (IDEX) that allowed the co-op to remain solvent. The co-op purchased modern honey storage equipment and imported beekeeping items for sale to members. As Africanization proceeded however, there were no customers for the bee equipment and little honey to process. The co-op kept afloat by selling agricultural supplies and groceries. Just as beekeepers had to diversify with Africanization, the support industry has had to do likewise!

Brazil

The Africanized bee story began in Brazil, the largest country in South America (8,512,000 km^2 – 3,285,632 mi^2)—fifth largest in the world. It has a population in excess of 190 million. Nearly 90% of the population lives on 10% of the land, mostly in the southern portion. It is largely tropical but the southern portion, particularly above 900 m (3000 ft), is temperate in climate.

The first beekeeping in Brazil, as in the rest of the Americas, was with stingless bees. The German bee was introduced about 1839 and Italian bees around 1922. Most early Brazilian beekeepers using *Apis mellifera* were European immigrants or their descendants, the vast majority of which settled in the three southern states. There was only a small internal market for honey products so beekeeping did not progress, despite the potential of abundant honey plants and favorable climate. Brazil, with production of approximately 6500 tons, was ranked 27th in world production statistics in 1956.

It was the decision of the Ministry of Agriculture with beekeeper support to import African bees. Professor Nogueira-Neto's writing in the 1964 *Bee World* reported the spread of the Africanized bee and changeover of managed bee colonies to more defensive colonies. He cautioned on further importation of African bees into other areas based on the early experiences in Brazil. He related that "one usually knows when *adansonii* (*scutellata*) first appears in a region…as people begin to complain, domestic animals are killed, people are severely stung and swarms appear."

Besides natural distribution, the spread of Africanized bee was human-assisted according to Nogueira-Neto. He relates that one queen breeder in particular "would not believe his Italian bees had turned into Africans…and continued to sell his stock." The queen breeder thought perhaps the reason his bees were smaller was "due to radioactivity from nuclear tests."

In Brazil, especially during the military dictatorship of 1964–1970, sensationalism of the press created great difficulty for beekeepers in finding and keeping apiary sites. There were numerous studies and pursuit of "efforts to find the best way to develop apiculture with these bees." At the same time, the outside world picked up on the "intensely negative campaign…exacerbated by sensationalized accounts in local news media." The legend of the "killer bee" was born.

In their 1991 *The "African" Honey Bee* book, Drs. Gonçalves, Stort and De Jong said the 1972 NAS report was part of the negative campaign because "the report focused on the problems rather than the solutions being developed by Brazil's beekeepers and scientists." In a "response" to the NAS report, a three-part article published in the 1973–74 *American Bee Journal*, Dr. Gonçalves said "…the beekeepers of Brazil are slowly adapting themselves to the Africanized bees and that…[it] is no longer a reason for one to renounce beekeeping as a vocation or an avocation."

Extension specialist Wiese discussed changes in Brazil with introduction of the African bee in his publications and beekeeping text. He reports that the bee "brought to Brazilian apiculture a new dimension…the Brazilian victory over the African bees awakened world interest [toward bees] among scientists and apiculturists." The introduction "initially caused a storm" for the beekeepers, which "generated great divergence of opinion and even discord."

Some news stories probably did sensationalize the new bee. One incident reported the shutting down of the town El Carmen near the Bolivian border; one six-year-old died. Another widely reported incidence involved a school bus that ran off the road into an exposed Africanized bee nest sending 46 students to the hospital. The most seriously injured student had a broken leg, surely not an injury caused by stinging bees.

Beekeepers can easily visualize how stinging situations would make news and easily lead to exaggeration. We need only look at what happens in our own press when a truckload of bees overturns or a beekeeper transporting bees has a traffic accident. Surveys of Africanized bee newspaper accounts by USDA scientist Nunnamaker along with another by Drs. Bill Wilson and Adrian Stoner illustrate how the news media sensationalize and don't get the story right.

Figure 10–3. A typical campesino.

Selection of Less Defensive More Productive Bees in Brazil

Africanized bees had the positive effect to bring the beekeepers of Brazil together "to discuss the problems." About 150 beekeepers and scientists attended the First Brazilian National Congress of Beekeeping in Florianópolis, Santa Catarina state, in May 1970. One presentation was by Cosenza who showed Africanized bees, or a hybrid, to be better honey producers than Caucasian bees.

Table 10.2

Honey Production 1972 (1 month), Minas Gerais State, Brazil

	Type of Colony	kg	pounds
14	African colonies	9.14	20.1
8	African colonies	5.55	12.2
10	Caucasian colonies	0	0
14	African X Caucasian colomies	9.13	20.1
8	African X Caucasian colonies	5.69	12.5

Some beekeepers considered Africanized bees advantageous because the "bees displayed greater productivity." Professor Kerr, in his 1967 article in the *South African Bee Journal*, claimed the African bees and their hybrids "produced two or three times as much honey" compared to European bees. Portugal-Araúyo published a separate study showing they stored twice the amount of honey as Italian bees and four times the average of German bees.

Some Brazilian beekeepers, especially those using the German bee, were no strangers to defensive bees. By the late 1970s "a significant reduction in the defensiveness of Africanized bees" was found in studies in São Paulo according to the article by Drs. Gonçalves, Stort and De Jong in *The "African" Honey Bee* book. Tests demonstrated that the same bees were noticeably less defensive in São Paulo compared to the hotter, drier climate of the northeast of Brazil.

Woyke, a Polish scientist, studied the Africanized bee in Brazil while on sabbatical leave in São Paulo. He compared the defensiveness of colonies of Italian bees, African bees and hybrid bees developed in São Paulo. His data are presented below:

Table 10.3

Comparison of Defensiveness: Italian, African & Hybrid Bees. 1969. Brazil

	Italian	African	Hybrid
# sec to 1st sting	19.3	2.9	9.0
# sec until irritation	22.9	7.3	15.0
# of stings in gloves	0	39.3	1.4
Time (sec) to restore calm	149	1801	243
Distance (m) bee pursue	28.8	170.9	31.4

It is clear that the hybrid bees were easier to manipulate.

What Woyke termed "hybrids" were later called "selected" bees. Duay and De Jong compared selected bees with wild-caught bees and non-selected Africanized bees measuring honey production of the 3 groups from 17 August to 8 December, 1993. The data are shown in Table 10.4.

Table 10.4

	Number of Colonies	Avg. Honey Production (kg)
Selected Africanized bees	118	45.3
Wild Africanized bees	22	31.8
Non-selected Africanized bees	13	37.7

Selected colonies were daughter colonies produced from queens of colonies selected for desirable qualities. The increase in honey production by the selected queen colonies (42.5%) over wild-captured swarms was particularly noticeable.

However, on the other hand authors da Silva and da Silva critically examined the few records of honey yield prior to Africanization and compared them to studies since colonization. They unequivocally state that the "goal of improved honey productivity was not reached" despite claims to the contrary. They cite the change from rustic hive beekeeping to use of modern movable frame hives and modern management as reasons for apparent yield improvement.

Success with Africanized Bees

In the three most southern states, those 10% of beekeepers remaining with bees were better prepared to handle Africanized bees as the climate of this area meant less defensive bees according to Dr. Charles Michener, one of the NAS team members to visit Brazil in 1972. The general public also adapted to the new defensive bee. Over 200 people reportedly have died from bee stings after introduction of the African bee but this number is no larger relative to total population than the figures given for the number of persons who die annually from bee stings in the U.S.

Beekeeping with Africanized bees enjoys some degree of success today in Brazil. The queen breeding effort at São Paulo has selected manageable bees within the Africanized bee population and this selected stock has been distributed widely. Beekeepers, too, have selected more manageable Africanized bees, which, along with management changes, has meant that beekeepers have "recovered" from the Africanized bee.

Dave De Jong, in a 1994 *Bee Culture* article, reported that "Although initially this bee created severe problems, as the beekeepers had no idea of how to handle them, the necessary techniques were developed, and now one can do anything with Africanized bees that can be done with European honey bees." He concluded his article by stating that the "majority (of Brazilian beekeepers) are fully satisfied" with Africanized bees. Scott Camazine and Roger Morse in a 1988 *American Scientist* article unequivocally stataed that "beekeeping is a thriving industry…properly managed Africanized colonies were excellent honey producers in Brazil's tropical and subtropical environment." Gene Robinson,University of Illinois bee specialist, opinioned that "success has to do with the fact the apiculturists and beekeepers have learned how to manage Africanized bees productively."

What has not worked in Brazil are programs that relied on or attempted reintroduction of European-race bees. An early extension program distributed over 23,000 virgin Italian queens over a 10-year period. It was "well supported" by the beekeepers but "the effort failed [since]...beekeepers ultimately killed such queens in spite of their relatively gentle progeny, because they produced less honey than Brazilian bees" according to the NAS 1972 report.

Orley Taylor found European bees not to be the answer in Brazil as there was "no way for beekeepers to maintain the status quo as Africanized bees move into an area." Dave De Jong asserts "return to the European bee was impossible because the Africanized bees were so superior in their adaptation to the environment that the European colonies practically died of hunger due to the competition for food."

In a 1996 visit to Brazil, University of Florida apiculturist Tom Sanford interviewed Dr. Antonio Carlos Stort, renowned University of São Paulo bee researcher studying Africanized bee defensive behavior. Stort asserted there is a "great variability in defensiveness...due to both climatic factors and production of alarm pheromone." Africanized bees are preferred according to Dr. Stort as "AHB workers fly faster, spend less time in the colony and more in the field [and] recruit sisters faster than Europeans...the Africanized bee is a productive and important part of the environment...[and] Brazilians...have gotten used to Africanized bees."

Much research, adaption by beekeepers and management changes, partially resulting from discussions at national congresses beginning in 1970, have "led to full development of Brazilian apiculture in terms of productivity" according to Drs. Gonçalves, Stort and De Jong. In 1972, average production was 15.8 kg/hive (35 lb) from an estimated 320,000 colonies. In 1988 it was 18.2 kg/hive (40 lbs) from approximately 2 million colonies.

Some production figures and hive numbers are condensed below:

Table 10.5

Year	Production Metric Tons	Number Of Hives
1957	6,527	—
1960	7,539	—
1965	7,904	—
1970	6,313	—
1972	5,052	320,000
1975	5,490	420,000
1980	19,000	1,100,000
1985	28,000	1,850,000
1988	36,000	1,980,000

Gonçalves, Stort and De Jong acknowledged this as "a very low figure." In fact, Brazil remains a country with one of the lowest average honey production figures in the world despite enormous size and potential. Production averages in the three southern provinces are better than for Brazil as a whole where with colony migration, mean honey yields of 45–54 kg/hive/yr (100–200 lb) are regularly realized with Africanized bees.

USDA scientists Tom Rinderer and Anita Collins compared the development of beekeeping in Brazil and Mexico, both countries with tropical and subtropical climates and beekeeping expansion in the past 30 years. Mexico, many magnitudes smaller in size (1,972,547 km^2 – 761,403 mi^2) and population (99 million), has "increase[d] its honey production several times more than the Brazilian increases" without the Africanized bee. USDA estimates were 2 million colonies and 38,000 beekeepers in Mexico in 1975; average production was 30 kg/col (66 lb/col). The FAO estimate for Mexico in 1984 was 2.69 million colonies with a 25 kg/col (55 lb) average before the Africanized bee arrived.

It is evident that beekeeping was severely depressed and the beekeeping industry of Brazil went through a difficult period of 15–20 years following introduction of the African bee. Beekeeping conditions have improved the last 10–15 years. Mark Winston, in his *Killer Bees* book, points out:

> "The Brazilian perspective was, and still is that the Africanized bee problem has been exaggerated and that properly selected Africanized bees are superior for tropical conditions.... Unfortunately, other Latin countries do not have the extensive expertise in bee genetics and beekeeping that are found in Brazil, and the ability of Brazilian beekeepers to respond to the Africanized bees has not been duplicated elsewhere in Latin America."

Peru

Peru is a large Andean country (1,285,215 km^2 – 496,093 mi^2) almost twice the size of Texas and slightly larger than Bolivia. With a growing population over 30 million living on a mere 2% of the land suitable for agriculture, there is a high population density in some areas and few inhabitants in others. Peru, like Bolivia, is ⅔ Amazonian tropics.

Robert Kent, a geographer at Kent State and later in California, conducted a survey of 112 beekeepers at 11 sites in Peru in 1981 and 1982. Five sites were Africanized and six were at locations still with European-race bees. He was able to document "consistent differences between the region affected by the Africanized honey bee and the region free of the hybrid (Africanized bee)."

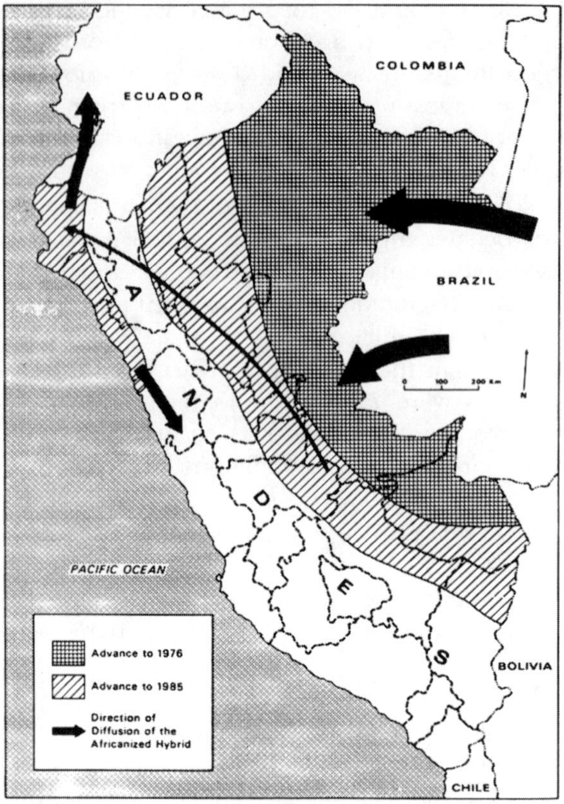

Figure 10–4. Disbursement of Africanized honey bees in Peru. (From R. Kent 1991 The "African" Honey Bee*)*

He found beekeepers better equipped, more progressive, better educated and using more modern equipment in the Africanized bee regions. There were more commercial beekeepers and fewer hobbyists where Africanized bees had spread (47% commercial in Africanized bee areas versus 95% hobbyist in areas with European bees). Kent reported the "average colony production is consistently greater" for beekeepers in the Africanized region, 30.5 kg versus 16.8 kg (67 vs. 37 lb).

At the time of his study, Africanized bees were distributed in the eastern part of Peru from the Amazon basin up to about 1500 m (5000 ft) elevation. Africanized bees had appeared in most of the five sites, designated in his study as Africanized, within the previous five years.

Half of his survey respondents said they had not noticed "any change in behavior of their bees." This finding agrees with what Marla Spivak found for beekeepers located at higher elevations in Costa Rica in her study of beekeepers during the first stages of Africanization in that country (Chapter 8). Beekeepers in the high elevations of Bolivia similarly could not recall when Africanized bees arrived as opposed to tropical

Bolivian beekeepers, where arrival of the Africanized bee was highly disruptive (Chapter 9).

Two of the five sites surveyed by Kent were on the Pacific coast of Peru where Africanized bee distribution was generally considered ("with a fair degree of confidence") human-assisted. Africanized bee genetic material was found in 300 colonies purchased in La Merced on the Eastern slope of the Andes, which were moved to the coastal area by two commercial beekeepers. The beekeepers claim they "requeened those [colonies] demonstrating aggressive characteristics" with Italian queens purchased in the U.S. after moving to the coast.

Kent concluded from his survey that while beekeepers considered the Africanized bee "bothersome...it is not a major problem." Some ⅔ cited the major advantage of the bee was that "they work harder and produce more;" 80% cited the major disadvantage as their "aggressiveness." Beekeepers reported stinging incidents with animals more commonly in areas with Africanized bees but human incidents were not statistically different in the two areas which "probably indicates that the public in the Africanized region is aware of the bee's defensiveness and consequently avoids the area around apiaries. Livestock...can make no such conscious efforts" so animal "accidents are frequent." Kent concluded that "Most beekeepers find management requirements more demanding and capital intensive with the Africanized bee."

Trinidad & Tobago

Trinidad & Tobago is a twin-island nation off the coast of Venezuela. Trinidad is 4829 km^2 (1864 mi^2—smaller than the state of Delaware) and lies about 11 km (7 mi) off the coast; Tobago is much smaller, 308 km^2 (119 mi^2) and a farther 30.5 km (19 mi) distance from Trinidad. Total population is about 1.3 million.

Trinidad has been repeatedly invaded by swarms of Africanized bees that have crossed the ocean from the South American mainland beginning in 1979. Tobago was free of Africanized bees until 1988 when beekeepers probably brought the bees to the island. After arrival of the Africanized bee, the number of beekeepers on the Island of Trinidad declined by ¼, (from 420 in 1979 to 324 in 1986). The number of colonies

Table 10.6
Beekeeping Statistics of Trinidad and Tobago[1]

	1978	1984	1986	1988	1992
# Beekeepers	419	373	324	430	430
# Colonies	7209	4910	4160	4807	4640
Honey prod. (liters)	90,833	55,237	43,056	64,895	83,520[2]

[1] Data courtesy of M.K.I. Hallim, Trinidad and Tobago Ministry of Food Production, Marine Exploitation, Forestry and Environment (MFPMEFE).

[2] Honey production from 1991.

fell from over 7000 to 4160 in 1986 with honey production reduced by over 50%. Since then the number of beekeepers has recovered but post-Africanization colony numbers still are 34% lower.

Honey production average yield is greater with Africanized bees. As shown in Table 10.6, the 1991 average yield of 25 kg (55 lb)/colony is greater compared to 1978 (16.8 kg – 37 lbs/col) before Africanization. It is evident, in colony yield and numbers of beekeepers, that the negative effects of Africanization have been turned around in Trinidad.

The Apiaries Unit of the Ministry of Food Production, Marine Exploitation, Forestry and Environment (MFPMEFE) began a large educational effort following Africanized bee arrival. They also established a 33-person bee "swat team," responsible for finding and destroying Africanized bee swarms. Since the initial capture of 12 swarms in 1979, this team has located and destroyed (or used in a queen breeding program) over 26,000 swarms/feral colonies. (Table 10.7)

Table 10.7

Africanized Bee Swarm/Colony Capture 1979–1991

Year	No. Swarms/ Colonies
1979	12
1980	50
1981	153
1982	613
1983	737
1984	3773
1985	2993
1986	2705
1987	3027
1988	2763
1989	3010
1990	3255
1991	3016

The Trinidad trapping program uses standard pheromone-baited traps. Large numbers of traps were placed along the beaches facing the South American mainland beginning in 1979. However, even with the trapping program, "all beekeepers' apiaries had become fully Africanized" by 1984. In one region, 130 deployed traps captured six swarms but the Apiaries Unit removed an additional 58 swarms from the area.

The Apiaries Unit used some of the swarms to start new beekeepers with hives and they used others to establish a queen breeding program. Queens were sold at subsidized prices (about US$2.25 for a mated queen in 1992). When European queens were imported, the colonies "hardly ever produced excess honey...[and]...quickly became depleted because they were outcompeted for bee forage by AHB." Unfortunately the program will not be continued; beekeepers are now encouraged to "select the least defensive and most productive" colonies to rear their own queens.

There are more beekeepers in Trinidad fifteen years after invasion compared to pre-invasion. Swarms represent a new resource for the island residents. An extension program, conducted by Mohamed Hallim, a Guelph University graduate, has trained over 700 new beekeepers. Hallim reports that although "the older beekeepers are slow in changing" to Africanized bee management, the two islands could support "at least another 50,000 colonies under sensible management." In addition to short courses and individual training, the Apiaries Unit uses a Ministry newsletter and several excellent extension publications on beekeeping to inform and instruct beekeepers. (Figure 10–5)

Figure 10–5. Trinidad extension brochure cover.

Honey and pollen come from a wide variety of floral sources in Trinidad and Tobago. Bees produce surplus honey during the drier months of January to June. Citrus is a major source along with coffee and tonka beans.With market protection (honey importation has been banned since 1935), the local price for honey is over US$30/gal. Pollen also sells well, particularly among the Asian-Indian population of the island, which is about 45% of the total racial mix of the islands.

The Africanized honey bee has led to an increase in stinging incidences. "Over 5300" human incidences have been recorded since Africanized bee colonization. There have been 12 fatalities—a rate of one person per year. Extensive searching of records by Hallim documented two known human deaths attributed to bees in the period 1903–1978. "Over 800" animals have been stung to death.

Recovery of the beekeeping industry on Trinidad and Tobago is attributed to the assistance program of the Ministry which includes capture and destruction of large numbers of feral swarms and colonies, along with a distribution of queens from their breeding program. The effective and readily available beekeeping extension program has trained many new beekeepers and retrained those willing to adapt to supply a highly profitable internal bee product market.

Venezuela

Venezuela is the wealthiest country in South America due to rich petroleum deposits. As elsewhere in South and Central America, the contrast between rich and poor is extreme. The population, growing at a rate of 2.6%, is about 14 million. In size (912,288 km^2 – 352,143 mi^2), it is larger than Texas and Oklahoma combined.

Spread of the Africanized bee to Venezuela occurred between 1975 and 1980. The Africanized bee advanced from south and east to west at 320–400 km (200–250 mi)/year. As in Panama, Venezuela had a growing beekeeping industry based on the European (primarily Italian) bee. Most of the large-scale beekeeping was concentrated in western Venezuela on the Caribbean coast from Caracas to the border with Colombia.

Prior to the threat of Africanized bees, Venezuela did not have beekeeping specialists in the Ministry of Agriculture. Spurred by a visit of Marshall Levin and Orley Taylor, the Venezuelan government increased their assistance to beekeepers. Individuals were encouraged to expand and buy new equipment at a favorable loan rate. Campesinos were offered incentives to begin beekeeping as part of a "protected belt" against Africanized bee invasion in the south of Venezuela bordering Brazil. An estimated 50,000 colonies, most managed by hobbyists many in primitive hives, were present before Africanization.

As the Africanized bee invaded Venezuela, much discussion arose about how the bee was different in southern Brazil from the bee moving along the coastal regions of northern South America. It was felt that the bee was "better adapted to the drier habitats" and moved more quickly along the coast than it did in the hot, humid Amazonian jungle. The bee however did enter and spread in Venezuela both along the coast and through the jungle.

As elsewhere, the Africanized bee brought many changes to Venezuela. Like Panama, the industry peak was the year following arrival of the bee into the country. Two commercial beekeepers, of 9 total, quit beekeeping and those continuing with bees reduced their operations by 50–75%. (In another version only two of 18 commercial beekeepers remained in business.) Less than 10% of hobbyists continued keeping colonies. Annual production of honey fell from 580 metric tons (of which more than half was exported) to only 78 metric tons/year. USDA scientists Rick Hellmich and Tom Rinderer stated "This near-collapse of the Venezuelan bee industry was due solely to the Africanized bee."

Forty-two U.S. beekeepers toured Venezuela in late 1984 and described the bee problem as "more severe than they had imagined." U.S. commercial beekeeper Keith Benson, writing about the trip, said "We found that the Africanized bees were not as bad as previously represented, they were worse!" Based on what he saw, Benson predicted "The invasion of the U.S. by the Africanized bee will be a major irrevocable disaster to agriculture, beekeeping and the general public." The group drafted a resolution soliciting support for research and action on "ways to retard or prevent the Africanized bee invasion" of the U.S.

In 1982 the USDA established a bee lab in Sarare, in the best beekeeping region of Venezuela. By 1985, significant research had already begun to define the biology and management of Africanized bees, especially with studies of defensiveness, swarming/absconding, honey production and pollination.

Venezuelan beekeepers faced with Africanized honey bees emphasize a time-demand change that includes everything from the time needed to prepare to enter the apiary, to better prepare the hives and apiary sites, to the additional time needed to conduct hive inspections. Africanized honey bees "require more inspections than are typically used in commercial beekeeping with European honey bees" according to Extension Specialist Montesinos. Based on the Venezuela experience he predicts the "Africanized honey bee population will be a challenge for American (U.S.) beekeepers."

A new generation of better-trained and educated beekeepers in Africanized honey bee management has developed in Venezuela. Commercial beekeepers are reporting 45–50 kg/hive (100–110 lb/hive) averages compared to 75–125 kg (165–275 lb/hive) before Africanization. Honey production in 1985 was 480 metric tons. The managed colonies now must compete with an estimated 1–2 million feral colonies.

One positive outcome for the beekeeper is that honey prices ($3.40/kg or $1.55/lb) have remained above the world market price in Venezuela. This is due partially to the fact that the national industry is protected by a tariff on imported honey.

Stingıng Incidents

There is reasonable data from Venezuela regarding the impact of Africanized bees on human health. During the initial 15 years (1975 to 1990), about 400 individuals died from bee stings. This is approximately 2.1 deaths/year/million people compared to the U.S. where annual death rate from honey bee stings is 0.07 deaths/year/million persons. However an estimated 100 persons in Venezuela (¼ of the 400 total) died in 1978, the year of greatest impact. The rate of 12 individual deaths 10 years later in 1988, about one person/year/million inhabitants, was the same as recorded the year prior to Africanization and similar to reports from Panama, Costa Rica, Trinidad and many other countries now inhabited by Africanized bees.

As in most other countries, flyers and extension materials were prepared in Venezuela to help beekeepers prepare and adapt to Africanized bee arrival. Beekeepers were advised to wear proper clothing, change apiary sites and adapt methods to manage the colonies to obtain a suitable honey crop and avoid incidents. There was also a coordinated program to educate the general public on the hazards of Africanized bees.

✧✧✧✧✧✧✧✧✧

Each of the five countries discussed here have been invaded by Africanized honey bees. Each has a different story of beekeeper adjustment and each has different governmental programs to deal with the bee. No one approach has been completely successful. As stated by Tom Rinderer, "Each new country…requires a lengthy period of adjustment during which beekeeping is first generally destroyed and then rebuilt."

A healthy beekeeping industry needs governmental action and funding to help beekeepers cope with Africanized bees. Not every country has access to the financial resources or the support available in Brazil. Availability of resources and such support leads to faster adaptation and a reduced negative impact of Africanization.

11

Making A Living with Africanized Bees

Defensiveness[1] is an easily-noticed difference among racial stocks of the honey bee. The expression of this behavior has significant impact on the beekeeper, beekeeping in general and with the general public's interaction with bees. Yet we still do not know much about this basic behavior of honey bees.

Protection of the nest, by exploring a movement that does not resemble a bee in flight or stinging a hive intruder after smelling the alarm chemical, are normal defensive behaviors—behaviors in response to a stimulus. Movement, jarring of the hive, unique smell, flight behavior and other features are some of the stimuli that might release defensive behaviors. Defense is not a hunting behavior—a search for something to attack—but a response to specific stimuli.

In European bees, the threshold of defensive response is high. We may term such bees quiet, docile, gentle, not aggressive, less defensive, or use similar words. Upon perception of the alarm odor docile bees merely explore or do nothing . In general, tropical bees, such as the Africanized bee of the Americas and the African bee of Africa, react more quickly to disturbances, sting more and pursue attackers longer and farther. Such bees are called mean, bad-tempered, aggressive, highly defensive, vicious, or similar terms.

USDA researcher Tom Rinderer found Africanized bees in Venezuela, when compared to African bees of Africa in behavioral assays, to "fall within the range of variation" in most behaviors, except they represented an "extreme" in their rapid defensive response to disturbance. Norm Gary notes that "extreme variation" in defensive behaviors is "typical [and] stinging episodes are highly probable, yet not always predicable" in Africanized bees.

Guzman-Novoa, while pursuing his PhD with Rob Page at the University of California, Davis, was able to reduce defensiveness of Africanized colonies to levels similar to European bees after only two generations of backcrossing Africanized queens to European drones. He has been able to demonstrate that worker defensive response in European-race colonies increased with higher proportions of hybrid workers.

---

[1] I prefer to use "defensive" or "colony defense" rather than "aggressive" to describe bee behavior. Many authors, beekeepers, and especially the general public, often use "aggressive." I believe aggressiveness implies a decision-based response which is not accurate. When bees attempt to sting they are responding to a stimulus or situation.

Testing for Defensiveness

No one single technique has been used as a standard measure of defensive behavior. The NAS committee members on their 1972 visit in Brazil modified a test developed earlier by Dr. Stort that quantified defensiveness by counting the number of stings in a patch of dark suede leather moved up and down in front of the colony. Another bioassay involves alerting with human breath and counting stings in a velcro patch.

USDA scientist Anita Collins, using the black-suede-patch technique, was able to illustrate differences in Africanized and European-race bee defensiveness. Africanized bees showed significantly greater defensive behavior including responding 2.4 times faster to the odor of alarm pheromone and 30 times more rapidly to the moving target of black fabric. Eight times as many Africanized bees compared to European bees typically sting the target. (Table 11.1)

Table 11.1
Time of response (sec) and number of stings of European and Africanized bees to dark swede patch after hive is jolted.

	European (E)	African (A)	Ratio (A/E)
Response to alarm pheromone	13.1	5.4	2.4
Response to moving cloth	9.2	0.3	31.0
Number of stings in target	10.4	85.7	8.2

Dr. Collins was able to quantify the components of defensiveness by taking photos and counting the number of bees in front of a bee colony before and after it was alerted by jolting once with a metal ball. Africanized bees showed a greater response as shown in Table 11.2 from her study.

Table 11.2
Comparison of number of bees in front of colony after hive alerted.

	European (E)	African (A)	Ratio (A/E)
Before alert (control)	44.2	70.6	1.6
30 sec after alert	61.1	121.6	2.0
60 sec after alert	66.6	141.7	2.1
90 sec after alert	84.3	172.8	2.1

In one defensiveness study in Brazil, a black leather ball 2 cm (0.79 in) in diameter was held by a gloved experimenter. Africanized and European bees orient to and sting the ball; Africanized bees, but rarely European, attack the target plus also gloves of the researcher. In a one-minute "leather glove test" 49 bees stung the right glove of the person holding the thread suspending the leather ball before an Africanized colony; zero bees stung the glove during a similar test with a European colony. Defensiveness is a pronounced trait in Africanized bees.

Figure 11–1. Bee stings in gloves.

In another series of experiments, Dr. Collins tested bee colony reactions of Italian bees in Louisiana and Africanized bees in Venezuela. Temperature, humidity, and test conditions were approximately the same at both sites. The Africanized bees responded more quickly and in larger numbers whether colonies were large or small. Large Africanized colonies averaged 8.2 more stings in suede-leather targets while small colonies left an average of 5.9 more stings compared to European-race bees.

Like many biological processes, the manifestation of defensiveness approximates a bell curve. Climatic factors have been shown to modify the defensive response. Brandeburgo demonstrated that a bee colony in Ribeirão Preto, São Paulo state, in the southeast of Brazil, were four times less defensive compared to the same colony tested in the northeast, near Recife, Pernambuco state. Recife is hotter and wetter while São Paulo has less temperature and humidity variation.

José Villa compared defensive behaviors of both Africanized and European bees at 520 and 2450 m (1700 and 8000 ft) in Colombia. At the lower elevation, Africanized bees responded twice as fast and left eight times as many stings. At the higher elevation, the bee response was slower and they stung only half as much.

Comparing the defensive response of bees in Venezuela with tests done earlier in Brazil by Brandeburgo, Dr. Collins concluded that the migrating Africanized bee had not become milder through "hybridization." The Africanized bees in Venezuela deposited an average of 0.85 stings/cm^2

of target when the colony was disturbed versus rates of 0.32 and 0.26 stings/cm^2 in Brazil.

The most defensive colonies in a study by Gloria deGrandi-Hoffman, of the USDA Tucson bee lab, were those with Africanized honey bee patrilines (i.e., some drones were African). Colonies with African queens mated to European drones were the least defensive. These findings were similar to those reported by California scientists Guzman-Novoa and Page who found most of the workers responding to colony disturbances were hybrid Africanized honey bees. Apparently, the type of queen in a colony has a limiting effect on defensive behavior. What seems to be most important is the type of drones that mate with the queen.

Toxic Reactions

One of the unique concerns with Africanized honey bees is that the ferocity of stinging incidents may result in a toxic reaction when a human or animal receives too many stings in too short a period of time. Mass envenomation can lead to direct poisoning of the body by the venom itself in contrast to an immune system reaction in an allergic response.

The venom of European and Africanized honey bees is not identical and because they are smaller, Africanized honey bees deliver a smaller amount of venom. However, studies have shown their lethality is equal and they are considered equivalent by medical professionals in determining bee venom sensitivity or to treat individuals for sting allergies.

Justin Schmidt of the USDA lab in Tucson has calculated that the LD_{50} of honey bee venom (2.8 mg/kg)[2] equilibrates to about 1100 stings/adult—that is half the population will survive 1100 stings. Fewer numbers of stings (100–300) have been known to kill humans. Older persons, because of other health problems, and youngsters, because of smaller body size, are at higher risk.

The total number of human deaths from Africanized bee stingings is not precisely known. In nearly 40 years of Africanization, perhaps 1000 individuals have died from bee sting incidents in the Americas. Venezuela seems to hold the dubious distinction of experiencing the majority of deaths—100 persons died in Venezuela the first year of colonization; a total of 400 in 15 years. Numbers from other countries, even Brazil, are considerably lower. I estimate that not more than one person per year has died in either Panama (2.5 million population) or Costa Rica (3 million population) from bee stings. In 10 years 8 people have died in the U.S.

Most stinging incidences with honey bees are not fatal. In the Panama Canal Zone for example, there were 35 stinging incidents during

---

[2]For comparison the LD_{50} for some common insecticides is: Parathion 3 mg/kg, Temik (aldicarb) 9 mg/kg, DDT 113 mg/kg, Rotenone 100 mg/kg and Sevin 500 mg/kg.

1985, the third year after Africanized bee arrival. For 1986–1989, there were 21, 22 and 16 annual stinging incidents respectively reported to Canal Zone health authorities. None were fatal.

The magnitude of Africanized bee incidents involving animals is not known. In conversations with beekeepers it appears to be a frequent occurrence. Brazilian Extension Specialist Wiese said "dozens of people and thousands of animals" have died as a result of Africanized bee stinging incidences in Brazil. Animals tied, confined in corrals or otherwise unable to flee are at greatest risk since they have no opportunity to escape once stinging begins. Smaller animals like chickens and dogs can tolerate fewer stings due to smaller body size. Their typical behavior of becoming agitated and "running in circles" additionally may lead to multiple stings. In Arizona, dogs and horses are stung more than other pets.

The location of the sting may also be a health-related factor in mass stinging accidents. Justin Schmidt has examined three cases of dog fatalities from sting attacks. In one, a German shepherd in Arizona, died after receiving over 3300 stings during a 45-minute absence of its owner. The highest density of stings was found on the dog's anterior muzzle and eyelids: 8 stings/cm^2 (52/in^2) compared to 0.01/cm^2 (0.07/in^2) on the feet/lower legs. Dark body fur patches had twice as many stings as light areas. It appears bees direct stinging to dark color and breath-odor clues so the face and exposed body parts receive disproportionally higher numbers of stings in dogs and, by inference, humans.

Reducing Defensive Behaviors

In numerous studies on defensive behavior of honey bees, Dr. Collins pointed out that genetic as well as environmental factors may influence the degree of defensiveness of a specific colony. In the dry tropics of Africa nectar resources are less predicable and depend on a non-seasonal rain pattern. It can be argued, therefore, that resources already harvested are more valuable due to the unpredictability of there being more nectar available to replace honey and pollen reserves. In Africanized bees defensiveness increases in the presence of large amounts of nectar. For temperate bees, the end of the nectar flow notably increases nest defense; they are less defensive during collection.

The Africanized bee tends to defend outside the nest. This is thought to be an adaptation to the tropical climate where vigorous defense within the hive itself might quickly raise the temperature and kill brood. Bees are more defensive when temperatures are higher—they react more quickly and are more likely to sting a moving target.

Brandeburgo, in his studies in Brazil, found defensive behavior more pronounced when colonies had more field bees, had higher numbers of guard bees, colony temperature was higher, and under climatic conditions of full sunlight and wind. The three factors, all interrelated, that yielded less defensive behavior were rainy weather, precipitation and lower atmospheric pressure.

David Fletcher in his chapter of *The "African" Honey Bee* book lists several stimuli that may lead to a greater stinging response by the African bee in Africa. He says that none "are unique to *scutellata*" but rather African *scutellata* have "a lower threshold of response" than bees of European origin. Once alerted the bees may "attack any human that happens to be in the vicinity." His list includes:

(1) high temperatures

(2) robbing by other bees

(3) bloom of certain plants (eucalyptus and aloes)

(4) strong odors of animals and certain plants (dogs, farm animals, sweaty beekeepers, some weeds when cut or crushed)

(5) dark coloration of objects

(6) alarm pheromone

One experience I had in Belize demonstrates how powerfully Africanized bees respond to alarm pheromone. Seeking to collect a sample of bees to dissect for tracheal mites I visited an apiary within a forested area late in the day when it was almost too dark to see to walk. I neglected to bring a jar to collect my sample so I simply wrapped 100–200 bees in the collecting net. From the characteristic odor, it was evident that the bees in the net immediately released alarm odor. Bees began to exit every hive in the apiary, keying on the alarm-chemical bundle I was carrying in the net. I exited through the dark, dense jungle growth but was unable to get rid of the flying bees. Finally I was free of stinging bees by leaving the net behind and walking a short 2 to 3 meters (10 ft) away. I retrieved my net and bee sample the next afternoon. No tracheal mites were found in the sample.

Why Are Africanized Bees So Defensive?

It is often stated that interaction with predators has led to evolution of the defensive African bee. It is thought that honey bees may have interacted with humans in Africa for over 3 million years. Even continuing to the present, native humans practice destruct-harvest in eastern Africa. In addition to humans, feral colonies also have to contend with ants, wasps, ratels (honey badgers), and other animals that dig up ground nests or climb vegetation to destroy aerial nests. However it is probably too great a simplification to solely ascribe defensiveness to centuries of interaction with man and other predators

If such human/bee interactions promote defensiveness, why are not other races, also under some of the same predation pressure of man and other pests, not likewise defensive? The dark bee of higher-elevation

African mountains, *Apis mellifera monticola*, exists alongside *A. m. scutellata*, as does *A. m. adansonii* of western Africa. Both races are not nearly as defensive although equally skilled human honey hunters collect their honey and brood.

Humans advanced from honey hunters to beekeepers in the Mediterranean region and Europe sooner than in Africa. With beekeeper management, bee colonies that are overly defensive are likely to be destroyed. Such selection pressure over time favors less defensive bees. In Europe, Russia, and in North America, feisty German bees were replaced by more gentle, manageable Italian bees beginning in the mid-1800s.

Importations of gentle European race bees (primarily Italian) into Africa were "massive" according to David Fletcher. The native *scutellata* population of the Pretoria area (where Italian queens were imported and raised, along with supporting European-race drone populations for over 30 years) "appears to have remained unaffected by its prolonged exposure to imported genetic material."

Dr. Fletcher references an African queen breeder as saying 30–40% of European queens did not successfully mate because most did not return from mating flights. Such losses were not observed in the native African bee. Perhaps predation on queens during mating flights (along with the difficulty of raising European-race drones because of parasitism by African drones) is the type of predation that we should focus on to explain Africanized bee defensiveness!

Dealing with Defensive Colonies

A part of the key to success with Africanized bees is how the beekeeper hives his/her colonies and location of apiaries. Additionally, preparations for colony inspections and the inspection itself are also critical in dealing with defensiveness.

The Hive

The Langstroth hive is the standard hive used throughout the world. Beekeepers in subtropical and tropical regions use one standard hive body to house the brood area during unfavorable (rainy or dry) weather. The second box is utilized for honey production.

The tropical beekeeper often contracts his hive building to a local carpenter who seldom understands the concept or importance of bee space. Too frequently the wood is not cured adequately. Frames may be twisted and too short or too long. Tropical beekeepers, like beekeepers everywhere, design or adopt equipment for use with their bees. But most Africanized beekeepers tend to streamline their operations, manipulating their colonies less using fewer extra items with their bees.

The Kenya top bar hive (KTBH) has been tried in Central and South America as an alternative to the Langstroth hive, in part because

Peace Corps reference materials recommend it and provide details on KTBH construction. (Figure 11–2) Top bar hives represent an improvement over rustic hives, but the effort has not been made to teach beekeepers how to use them in the Americas. Tom Rinderer found Africanized bee colonies maintained in top bar hives to be smaller in population and to produce less honey compared to colonies kept in Langstroth hives.

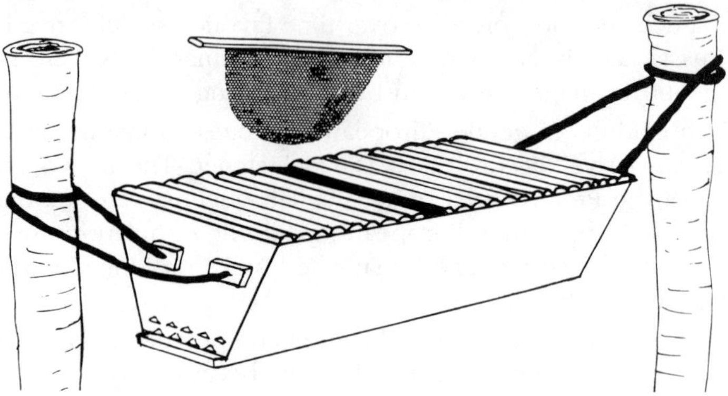

Figure 11–2. Kenya top bar hive. (Source IBRA)

Rustic Hives

Beekeeping with traditional or rustic hives, particularly in development programs, decreases after Africanized bees colonize an area. However, in some regions, rustic hives or colonies with non-removable frames remain in common use. A recent estimate suggested that as many as 60% of the bee hives in the Cochabamba valley of Bolivia are still rustic or primitive. The bees are all Africanized, but are less defensive than in tropical regions so beekeepers persist with primitive hives.

Many beekeepers at higher elevations in Bolivia still keep their hives close to the home and in areas where there are other animals to help protect the hives from being stolen or vandalized. Cochabamba is very dry and the native vegetation has been degraded so there is little means of concealing an apiary. The best hive security is the ancient one—locate the colonies where you live or work.

In the Cochabamba valley, I visited with José Rodriguez who has kept honey bees in rustic hives for over 25 years. His 70 colonies are kept around his home along the main Cochabamba–La Paz highway; about 10 colonies are in modern, movable frame hives. He says he "prefers the rustic hives over the modern ones" and believes he can harvest a greater amount of honey from the rustic colonies.

His standard rustic hive is a wooden fruit crate. The box is about 24 cm by 30 cm by about 60 cm high (9 x 12 x 24 inches). Inside, there are two or three wooden horizontal braces about half way up the box that serve to separate the hive so comb with honey is above and the brood cluster area is below. This helps separate brood and honey. Notches are cut in one bottom side for an entrance and an alighting board is frequently added below the notches. The boxes sit off the ground on cement blocks, wood or half a rusty honey drum or other convenient supports. (Figure 11–3)

Figure 11–3. Rustic hives in Bolivia.

During a November afternoon (spring in Bolivia) I watched José harvest two of his rustic hives. It was the first of what hopefully would be three harvests for the season. The hives were strong and heavy with honey. The harvest was amazingly rapid—I was surprised at the amount of honey obtained from each hive.

To harvest, the hive is transferred from its stand to a table and inverted. An empty hive is put in the place of the removed hive to capture the returning foragers. In addition to locating the harvest table downwind of a smoldering fire José used a smoker as well. The one unattached side of the hive is first removed and the colony smoked from the bottom to drive the bees from the honey storage comb to the brood area (since the box is reversed). Then using a series of long bladed knifes and spatulas, pieces of comb containing honey are removed from below and put into a plastic colander.

The comb pieces are cut in irregular shapes roughly 8 by 10 cm (3 by 4 in wide). A homemade knife with a 1-inch right angle at the tip is used to cut upward between parallel honey-filled combs. The comb is cut in sections with the front half removed before the back portion.

José's daughter brushed the cut pieces of comb of any remaining bees and transferred the pieces into a covered 5-gallon (19 l) plastic container. Pieces of comb that had brood or pollen were separated from portions with capped honey and placed into another bucket. There was no separation of capped from uncapped comb but there were surprisingly few uncapped honey cells.

The entire harvest took no more than 20 minutes during which time the upper 30 cm (12 in) of honey-filled comb was removed from the hive and placed in the covered bucket. The total harvest filled most of the 5-gallon container, an estimated 20–25 kg (44–55 lb). It would be sold as harvested in plastic bags like cut-comb honey.

When most of the honey-filled combs had been removed, the hive wall panel was replaced. The hive was lifted from the table, reinverted to its correct orientation, and moved back to the original location. The heavy beard of bees on the sides and top was gently brushed onto the ground after the hive was placed back on the hive stand. A piece of wood was positioned from ground to entrance for the bees to walk back into their hive. The bees readily reentered to cover the brood combs, left intact during the honey harvest.

Both the beekeeper removing the pieces of comb and his daughter brushing bees from the pieces of comb were stung several times on the hands. However, use of smoke and disoriented bees meant that stings were not a problem. A bucket of clean water was used to keep the brush and knives clean and wash accumulated honey from the hands of the two persons harvesting the honey. Harvest was messy but not unusually so.

In an hour, José harvested nearly 45 kg (100 lb) of honey from two rustic hives. He generally harvests two or three hives toward the end of the day and then sells what he can the next morning. He said he easily sells all his harvest from his 50–60 rustic colonies. He preferred this harvest method over using the extractor which he used only to harvest his standard movable-comb hives.

All of this was accomplished with Africanized bees!

The Apiary Site

In the tradition of European beekeeping, bee colonies in the Americas were kept close to houses with other domestic animals. Apiary sites open to sunlight meant constantly keeping vegetation down so hives are usually located close together, several to a hive stand. (Figure 6–2, page 68) Apiaries in Africa, in contrast, are located away from homes. African hives are most frequently hung in trees for protection against ants and small mammal pests. (Figure 3–3, page 32)

After arrival of Africanized bees, beekeepers in South and Central America found it necessary to move bee hives and apiaries away from people and animals. It is easy to discuss the ideal apiary site for the Africanized bee with new beekeepers; established beekeepers are slower to grasp the importance of using more isolated apiary sites. With initial colonization, a large number of colonies are simply abandoned as alternative sites are not always available for the beekeeper.

Various recommendations have been published for apiary sites. (Figure 11–4) Most recommend placing colonies on individual hive stands 1–2 m (3–10 ft) apart. Apiaries should be located 200–500 m (660–1650 ft) from residences, roadways, public access areas or areas where animals occur. Almost all recommend a reduced number of colonies, both because of colony defensiveness and for the intense competition from feral colonies.

Isolation of the apiary site is required but the concept should be interpreted as isolation of the bees. A vegetation barrier, fence or isolating structure serves just as well as distance. In my experience, distance is not important in areas of heavy vegetation growth as much as isolation. An ideal site is a corral inside a ring of thick vegetation that isolates the bees from view and accidental entry into the apiary. A heavy vegetation barrier causes the bees to fly up and above the heads of passers-by.

Figure 11–4 Panama Peace Corp illustration of ideal apiary (at back with vegetation barrier and warning signs 200+ m from road/people/animals).

Maintaining isolation is important. It is especially important not to trail angry bees out of the apiary following colony manipulation. For this reason, I recommend two entry/exits where vegetation is dense. For the initial entry, an opening sufficiently large and clear of vegetation to permit easy entrance and movement of equipment into the apiary is helpful. To exit, a trail through the thickest vegetation will help the beekeeper get rid of the halo of bees.

Ground movement and vibration are disturbances that lead to stinging accidents. The correlation of stinging incidents with use of heavy equipment such as lawn mowers/trimmers, traffic, and horse passage, is not accidental. Many instances exist of heavy, odor-belching, earth-shaking pieces of equipment disturbing a bee colony and then innocent persons having to fend off and avoid angry bees for the rest of the day.

In Panama I visited an apiary site near a dumpsite. The beekeeper complained that the bees were very defensive and that they were stinging the men on the garbage trucks. I could feel the ground vibrations from passing trucks, and garbage odors were very strong in the apiary. The beekeeper reluctantly agreed to move the apiary when I suggested that there was little he could do to prevent stingings and that he would lose the bees sooner or later if the workers decided to "remedy" the situation. After he moved the bees he said he could see a difference almost immediately; the bees were easier to manipulate at the new site.

Beekeepers everywhere report difficulty in finding suitable apiary sites after Africanization. In Brazil during the initial "sensationalism by the media," beekeepers had difficulty getting apiary sites "in commercial orchards and eucalyptus plantations." according to Drs. Gonçalves, Stort and De Jong. New beekeeper problems, e.g. stealing of honey and acts of vandalism, have increased since beekeepers now isolate apiary sites.

Apiaries still need to be protected from things like grass fires and floods. Protection from non-domesticated animals is less important in Asia and South America, though some beekeepers have learned to fence or screen colonies from hoofed animals. Colonies close to the ground are more susceptible to anteater and armadillo attack. Fire ants heavily predate on feral colonies close to the ground in Texas, according to Paul Jackson. Army ants are a universal predator throughout the tropics.

Hive Stands

The hive stand is an important consideration with Africanized bees. A sturdy hive stand about 30–50 cm (1–2 ft) high results in less bending and is easier on the back. In Bolivia and elsewhere, the common individual hive stand is made of metal concrete-reinforcing rod (5/8 in). Wooden hive stands are of little value since termites and rot quickly destroy them. Other non-wood materials such as adobe or cement blocks, bricks, metal angle iron, old metal barrels, used tires, etc., are also used.

Beekeepers have found it helpful to separate individual colonies by 2 m (6 ft) or more. (Figure 11–5) Common hive stands transmit vibrations from one colony to the next and release of alarm odor alerts colonies not being inspected. An elevated hive stand assists with minimizing body movements. Elevating the bees off the ground means less bending to expose ankles or straining of protective tape used to secure holes in clothing, pants to boots, shirt sleeves to gloves, and other openings.

Figure 11–5. Hives on individual hive stands .

Elevation of colonies aids in reducing vegetation and improves air circulation beneath the colony, both of which are additionally useful for pest control. Elevated hive stands are important, too, in the constant battle against termites, ants and weeds in the tropics. Tropical beekeepers elevate colonies to permit use of barriers against these pests. Used motor oil, asphalt mixed with oil, or other repellent materials are used to protect the colony. In tropical areas, beekeepers seek to continually clear vegetation in the apiary to help detect and reduce pest entry.

The Africanized bee is extremely sensitive to movement and vibration. Beekeepers have learned to use gentle, deliberate movements around Africanized bees. Rough handling, bumping and especially failure to handle frames smoothly, all provide stimuli to the bees to sting. The Africanized bee uses more propolis and prying is always necessary to free frames and remove hive bodies. Lifting supers, moving and inspecting frames without vibration and bumping helps avoid release of defensive behavior.

Personal Protection

Safety should not be left to chance when entering into and working in an apiary with Africanized bees. Managing defensive Africanized bees is more costly and time-consuming since protective clothing is more expensive and time is required to dress properly. Beekeepers or employees with little protection are not very useful in the apiary as they cannot work effectively.

Reliable coveralls or a bee-tight suit is recommended for Africanized bee management. (Figure 11–6) Veils, particularly, need to be sturdy and well-sealed. Clothing should be light-colored (white or, as many prefer, orange-colored coveralls) and smooth, not rough-textured. Double pants, heavy shirt, and plastic bags tied around the feet and hands serve as body coverings for less prosperous beekeepers. Bees still sting through clothing, particularly when sweat plasters the clothing against the body. Padding the shoulders and thighs is often helpful, as is use of two pairs of coveralls when bees are expected to be particularly defensive.

Figure 11–6. Beekeeper ready for work with veil, coveralls, gloves and boots.

Beekeepers with Africanized bees normally wear gloves, or at least keep a pair close by in case they are needed. Heavy leather gloves work well but are hot, cumbersome, expensive, and retain sting odors. Most beekeepers prefer cheaper light-colored plastic or rubber platex-type gloves. These are often worn over cloth gloves to reduce perspiration of the hands and permit easier removal. Bees are reluctant to sting rubber gloves and they do not retain the alarm chemical odor.

Smokers/Manipulation Cloths

Beekeepers frequently elect to use a large (Brazilian) smoker (30 x 15 cm firebox) when working Africanized bees (as illustrated on the cover). The smoker should be started and functioning properly before entering the apiary. Usually smoking is a full-time job for one person.

Manipulation cloths (Figure 11–7) are very helpful to keep flying bees within the colony and to reduce the chances of robbing. Some beekeepers simply use wet towels or cloths. The wet towels or a manipulation cloth are placed over the top bars of boxes that are opened and then rolled back as the beekeeper proceeds with the colony examination. Either can be a valuable tool with Africanized bee colony inspection when used correctly although they may initially be cumbersome.

Figure 11–7. Manipulating cloth (or wet towels) to cover open hive body may reduce stings.

Management for Honey Production

In Brazil Dave De Jong has found "rough handling and carelessness by beekeepers during manipulation of the frames and other parts of the hive, provoke a defensive reaction in the bees." He remembers an instance when a visiting U.S. researcher tried to staple a plastic entrance reducer onto a small colony without using smoke. The bees began to sting "with an intensity that I have never seen...in ten years of working with Africanized bees."

In another instance, Dave relates that an ORISA-sponsored tour of visitors from Central America and Mexico were present but "not really absorbing some of the many small but important modifications in management...necessary in order to control a colony of Africanized bees." As a consequence, when it came their turn "the first 'volunteer' proudly showed everyone what he could do. In less than a minute the bees had chased us out of the apiary...when someone has made a gross error (usually a foreigner), I have had experienced beekeepers say that's how the bees were in the old days."

Dave believes keys to success with Africanized bees are "in the apiaries (location selection), in the equipment of the beekeeper, and in the management and bee handling practices." To this we might also add selection of gentle bee stock and elimination of highly-defensive colonies.

There are some basic differences in bee-foraging behavior between the tropics and temperate areas. Tropical flowers tend to produce more dilute nectar and, as a resource, are more widely distributed. Temperate areas have larger concentrations of high-nectar-yielding plants. Native and introduced honey bee races have adapted to these two patterns.

Africanized bees use a higher amount of available comb space for both stores and brood—80% compared to 65% for European bees. When resources, especially pollen availability, improve, the colonies rapidly increase brood rearing. This means less space for workers to spread unripe nectar to take advantage of the warm, moist hive environment for honey ripening. Heavy brooding and reduced storage space result in lower harvests where the nectar flow is short and intense—it so happens that tropical flows are more likely to be long and less intense.

Figure 11-8. Panama beekeeper Juan Malivern with honey-filled frame.

From studies where bees from European colonies were placed into Africanized colonies and vice versa, Canadian researcher Mark Winston and his student, Katz, found that tropical bees begin to forage at an earlier age (20 versus 26 days). The European bees placed into Africanized colonies foraged at an earlier age than normal. They concluded that Africanized workers have a higher threshold of the stimuli that lead to foraging by adult bees so they become foragers at a younger age.

Tropical bees collect more nectar under weak nectar flow conditions than do temperate bees because tropical-evolved bees forage more as individuals, a behavior termed "gleaning." The supporting evidence is as follows:

Africanized bees:

- make shorter inspection trips
- make more round trips per hour.
- switch from a declining forage site more quickly.
- prefer flowers with higher nectar concentrations
- dance less and also follow dancers less.

Gleaning occurs in European bees, according to USDA scientists Tom Rinderer and Anita Collins, for example in conditions typical of autumn in temperature areas. "When better resources are available, they [Africanized bees] will forage, but with less intensity than European bees....Africanized bees are adapted to conditions where an individual gleaning type of foraging is more successful. European bees are adapted to conditions where intensive foraging reliant on dance communication is more successful."

Rinderer and Collins conclude that foraging behavior differences in Africanized and European bees "are rooted in the ecological differences" related to where the bees evolved. Unpredictable rainfall in Africa keys the process of African bee development, whereas in European bees, seasonal predictability in amount of available empty comb space, influenced by increasing day length, are the prominent foraging keys

Tom Seeley of Cornell considers that European honey bee foragers primarily use the motivation of dance recruitment as the stimulus to forage. Smithsonian tropical bee expert Dave Roubik says this is "probably maladaptive in the tropics." Under tropical conditions, increased individual scouting leads to faster discovery of a pollen/nectar source than waiting for instructions received from a dancing bee. The bee that finds flowers first is better able to exploit the source. Africanized bees are more opportunistic, relying less on dance language.

Pollen Collection

Africanized bees excel in one area—collection of pollen! Africanized bees collect and store larger amounts of pollen, partly because a greater proportion of foraging-age bees collect pollen versus nectar. European bees have a larger, nectar-only foraging population.

Pollen collection opportunities are enhanced where humans create disturbances in the jungle. For example, honey bees readily use corn and palm pollen, two common plants of disturbed areas. Undisturbed tropical forests characteristically have a wide diversity of flowering plants at any one time but also smaller numbers of each. Harvest must be done quickly because the weather will change, competitors will find the same source, or the few flowers will be quickly harvested and the source rapidly exhausted.

USDA scientist Bob Danka simulated intensive migratory management of 15 colonies of European and Africanized bees in Venezuela. The colonies were moved six times over a two-month period during the dry season. The Africanized colonies swarmed frequently, stored less honey and had smaller foraging populations than the European colonies, though they collected more pollen during their foraging forays. Incidences of stinging and running behavior hampered inspection and management three times as often in the Africanized bees in this test.

✧✧✧✧✧✧✧✧✧

Defensiveness in Africanized bees is legendary. A wide range of stimuli elicit stinging reactions in Africanized bees. Africanized bees react more rapidly and forcefully compared to European bees. With release of alarm chemicals, the Africanized bee may sting humans and/or animals enmass sometimes resulting in a toxic reaction. Exactly why Africanized bees are so defensive is not known.

Some Africanized bee colonies are gentle, just as some European colonies will readily sting but, in general, the average Africanized bee colony is defensive. Extreme examples of nest defense occur with some frequency posing a problem for the general public and agriculture, including animal care. Beekeepers adjust to defensive Africanized bees by changing or modifying apiary sites, using individual hive stands, donning protective clothing, utilizing large smokers and changing hive manipulations.

Although the bees have changed, incidents become less frequent after colonization largely because beekeepers learn to adjust to the bees. Beekeepers who do not adjust soon abandon beekeeping due to the extreme defensiveness of the Africanized honey bee.

The hoped-for superior honey production capability never materialized for Africanized bees. During heavy nectar flows, European bees collect more nectar than Africanized bees. Under poorer resource conditions, the Africanized bee is a better gleaner and more efficient forager. European bees tend to remain as unemployed foragers in the hive waiting for dancers to signal food availability while Africanized bees are less dependent upon this foraging stimulus.

Thus under limited conditions, the Africanized bee is a better bee for honey production and pollination of tropical plants. However over the total range where it has spread, it has caused significant changes to beekeepers and beekeeping.

12

The Tropical Africanized Bee

Perhaps nothing is more controversial about the Africanized bee than its honey-producing capabilities. Consider these four statements made in the 1970s, all about Africanized bees.

> Professor Portuga-Araújo of Brazil, in one of the earliest scientific reports available in English about Africanized bees in Brazil, stated that Africanized bees "stored between two and four times" the amount of nectar that European colonies stored in one honeyflow period in Brazil.

> "About 50% of them [beekeepers in São Paulo state, Brazil] prefer to work with Africanized bees because of their high productivity. Many beekeepers in Brazil do not care about stinging if their bees make money....Of course, since the introduction of the *adansonii* [*scutellata*] bees into Brazil most of the hobby beekeepers have gone out of business." (Dr. Lionel Gonçalves 1974)

> "Commercial beekeepers especially in southern Brazil mostly feel that the Brazilian [Africanized] bee is superior to any race they have had before.... Beekeepers value the large crops of honey that the Brazilian bee produces." (Dr. Charles Michener 1975)

> "Production tests between the bees of European races and the *Apis mellifera adansonii* [*scutellata*] bee revealed greater production with *adansonii* bees...the most productive apiculturists are satisfied with the production...the productive capacity of the *A. m. adansonii* bees...completely compensated fully for the difficulty faced by the Brazilian apiculturalists...the beekeepers learned quickly and preferred to confront more aggressive bees, since they provided good harvest of honey." (Wiese 1977)

And consider these statements made in the 1990s from researchers who investigated Africanized bees.

> "Unfortunately the African bees did not live up to their reputation for honey production (in the Americas), and many of the same characteristics that have proven so successful in the wild are highly undesirable for bee management....The reputation of Africanized bees as good honey producers in tropical habitats comes from their ability to collect more honey than European bees under conditions of poor, dispersed honeyflows." (Dr. Mark Winston 1992)

> "Although there are some reports from Brazil of increased honey production with Africanized bees, there are many more reports of decreased production, disrupted beekeeping practices and seriously depressed beekeeping industries." (Rinderer and Collins 1991)

> "Africanized honey bees are undesirable stock for highly productive beekeeping. Under the nectar flow conditions required for commercial honey production, Africanized colonies exhibit lower honey production than European colonies." (Pesante, et al 1992)

Early comparisons by Brazilian scientists Kerr, Cosanza and Portugal-Araújo (Chapter 10) all demonstrated two- to four-fold increases in honey production of selected Africanized bees compared to Italian or German bees in side-by-side tests. In a 1993 test of honey production in Brazil by Drs. Duay and De Jong, the selected Africanized bees produced nearly 19% more honey than the non-selected or wild-caught colonies and 76% more than the F_1. Indeed, Drs. Page and Kerr in *The "African" Honey Bee* book state Brazil has become a major honey-producing country with Africanized bees because "the Brazilians have put considerable effort into selecting usable African stock."

In a USDA-maintained apiary site in Venezuela, European bee colonies of similar size collected more nectar per day than Africanized bees during the dry season, the more productive portion of the annual cycle according to beekeepers. During the wet season, a minor nectar flow period, Africanized bees stored more honey than European bees.

These studies and numerous others clearly point out that it is necessary to clearly define "better honey producer" when discussing the Africanized bee. As Mike Allsopp, a bee researcher from South Africa, home of the African bee stock carried to Brazil, observed "The AHB

[Africanized honey bee] has shown better honey production in South America than the EHB [European honey bee] ever did."

The American Revolution

Beekeepers of North America began a revolution in beekeeping. The changes started in the mid-1850s when Langstroth patented a practicable movable-comb hive, based on his discovery of bee space, and wrote *The Hive and the Honey Bee* describing how to use his hive. Langstroth's movable-frame hive, along with many imitators, was readily adopted because it made beekeeping profitable. Human beekeepers rapidly advanced from honey robber to bee manager.

Humans have not changed honey bees as has been done for other animals in our care. Instead, a better bee was found. Langstroth liked the Italian bee. He, along with others, recommended their adoption once they were available from U.S. queen breeders after 1861. Since it was a better bee to manage in temperate climates progressive beekeepers quickly accepted it.

The movable-frame hive and adoption of Italian bees were quickly accompanied by other advances (foundation, honey extractor and bee smoker, for example) that led to increased honey harvests. For awhile, the industry split into two camps—Langstroth's small hive/comb honey faction and the large hive/extracted honey beekeepers. Eventually consumers came to accept the purity of liquid honey, aided by the first Pure Food and Drug law passed in 1906. Beekeepers further increased extracted honey production because they could produce and sell more products.

Bees of European origin, primarily the Italian bee, helped make change possible. What works in one region however is not necessarily the solution for another. We now realize that tropical bees cannot be managed by the same methods as used for the European bee. Tropical bees interact differently under human care.

In South and Central America, where previously only nests of the native stingless bee were harvested, arrival of Africanized bees has resulted in the availability of a new resource—abundant swarms and natural (feral) colonies of honey bees. Many native peoples, like cultures before them in Africa and Asia, have become honey hunters and beekeepers.

Beekeepers and the general public have learned to adapt to the Africanized bee. There is no other alternative. Africanized bees have not proven to be a public health hazard. Intensive attempts to remove swarms and feral colonies along the Panama Canal reduced abundant wild colonies/swarms following colonization but the bee continues to flourish. Problems eventually decrease after colonization when people learn to respect bees and leave them alone.

In underdeveloped countries, 50% or more of the population exist by subsistence agriculture. For some of these persons, the availability of abundant swarms and high numbers of feral colonies of Africanized bees represents a new and exploitable resource. Studies such as one conducted by Wilkins have demonstrated that traditional farmers (the campesino or rural subsistence farmer) "respond much like their industrialized counterparts: they identify goals, minimize costs (most of which are labor) and they maximize gains by careful use of available resources."

Part of the key to the training of campesinos is the application of appropriate technology. The appropriate technology for some native populations may not be to learn how to become beekeepers with Langstroth hives but rather how to be honey hunters. Canadian apiculturist Gordon Townsend described a project in the early 1970s in Paraguay in which teaching modern beekeeping, that included the importation of hundreds of Italian queens, did not succeed. However, an unexpected benefit was that the Indians became successful at collecting honey from the wild nests of recently-arrived Africanized bees and selling it at local markets.

Repeated efforts to introduce temperate-zone bees to the tropics have not met with success. Too much biology is out of sync with the environment—features that do not promote survival, even with care and management by humans. Those beekeepers who are willing to adapt and successfully practice post-Africanized-bee management survive and have the opportunity to prosper.

We have already learned some valuable lessons in our dealings with Africanized bees. We initially sought means to stop their spread northward. Thank goodness proposed "solutions" like an insecticide curtain were never implemented. Beekeeping suffers sufficiently from pesticides without creating a larger problem. A lot of money was spent monitoring Africanized bee progress but we lacked some important information to try to slow their movement through Central America. Time was not in our favor.

Each country invaded by Africanized bees learned the value of educating the public about the importance and necessity of honey bees. Several found ways to keep Africanized bees from becoming a serious problem, though not all countries could monitor their movement or attempt removal. Beekeepers learned to adapt or simply gave up. Many U.S. beekeepers learned about the new bee by visiting Central and South American beekeepers already coping with the Africanized bee.

The $64,000 Question

In European bees, the winter is a major feature of the annual cycle. In Africa, *A. mellifera scutellata* is described as a bee that inhabits hot, dry areas. However, it lives in areas where snow and freezing temperatures are common for part of the year. For example, Al Dietz found Africanized bees at 3300 m (10,800 ft) in Argentina. I examined Africanized bee colonies in La Paz, Bolivia, at over 3600 m (12,000 ft).

At Cordoba, Argentina, (1400 m – 4600 ft), Al Dietz and his students found no significant difference in survival, food consumption or subsequent weight gain between either large- or small-sized Africanized and European bee colonies. Follow-up studies confirmed Africanized and European colonies surviving up to three months at freezing temperatures. Africanized bees died first one year but there was no difference the second year between Africanized and European bee survival.

What do high-elevation studies tell us in terms of eventual distribution in the U.S.? Although the bees readily colonized tropical areas of the Americas, they dispersed upward in elevation less readily. In Costa Rica, Colombia and Bolivia, Africanized bees are less common and their highly defensive behavior less prevalent at higher elevations. The density of colonies is lower and rates of swarming and absconding are lower.

When intentionally carried to higher elevations, Africanized bees are able to survive and eventually grow populous enough to swarm. José Villa found Africanized bees colonizing artificial nest cavities at low densities up to 2750 m (9000 ft) in Colombia. In Bolivia, arrival of Africanized bees in the high-elevation valleys of the Andes Mountains was not as notable nor as distinct a contrast as it was in the lowland tropics. Over time however Africanized bees have completely colonized the high lands, though they are not nearly as defensive as in tropical conditions.

From her studies in Costa Rica, Marla Spivak identified several factors for reduced bee success at higher elevations, such as fewer resources, less successful queen replacement, population dwindling and colder temperatures limiting foraging. She found fewer differences between Africanized and European race bees at higher elevations (2800 m – 9200 ft) versus great differences at a lower elevation (900 m – 3000 ft). Africanized bees migrate to higher elevations, successfully live there through colder temperatures without flower resources and yet still swarm/abscond at such locations.

Overwintering ability figures into calculations on how far the Africanized bee will spread into the U.S. Summarizing the studies of his students in Argentina, Al Dietz concluded that colonies of Africanized and European honey bees "showed no significant differences in the rate of winter survival." He concluded winter survival may "not be the main factor in limiting the distribution of African bee genetic material in Argentina."

Mark Winston believes that "no single trait" will prevent Africanized bees from succeeding in temperate winters but "the combined characteristics that make these bees tropical will limit their spread northward." The features likely to be most limiting, according to Dr. Winston, will be small nest-cavity preference, frequent swarming and absconding, shorter life expectancy of Africanized bees, and the nervous nature of the bee resulting in a less compact cluster.

Orley Taylor predicted that pockets of Africanized bees will exist north of normal overwintering limits. How extensive and dominant these become rests on:

(1) how permanent and widespread they are,

(2) how many encounters there are between bees and humans,

(3) how many managed or feral European bees are in the area to permit hybridization.

As a generalization for the Americas, Taylor observed that "...wherever European bees have been used successfully, high-density populations of African[ized] bees have become established."

Environmental Factors

As we have come to better understand the Africanized bee and some of the environmental and human interactions to which it is exposed, we have developed a better appreciation of the differences between temperate versus tropical bees. What are some of the environmental influences that shape and change bees? In temperate climates there is an annual seasonal cycle, keyed to a predictable and easily-measured daylength, shaped by a variable-length winter when foraging and brood rearing ceases, bee life lengthens, bee behavior changes (they cluster) and the bees remain inside the hive. In temperate areas there is one major flow; the annual cycle shows one peak in brood and adult population. Skillful beekeepers know this cycle and manage colonies with anticipation rather than simply reacting.

Unlike temperate areas, where the dominant feature of bee biology is a predictable annual cycle, bees in the tropics are less influenced by seasons with varying day length. There is no winter in the tropics. Temperatures do not get low enough to necessitate clustering. Hot temperatures promote opposite behaviors like defending outside, rather than from within, the nest cavity. Bees forage early and late in the day rather than mid-day. The active season is stretched out over a longer period of time, keyed to wet and dry seasons. When tropical beekeepers speak of "winter" they are referring to the wet season.

In a few tropical areas (Africa, Guanacaste in Costa Rica, and the Andean altiplano), the dry season is the dearth period. Without rainfall, everything simply dries up. Rains, when they do arrive, transform the dry, parched earth into a beautiful, abundant nectar/pollen flowering landscape.

The key to understanding the success of the Africanized honey bee is to recognize that it evolved in the tropics. Africanized bees do better in this habitat than temperate-evolved bees.

Seasonal patterns in rearing of brood, swarming and absconding behavior, and even length of adult worker life, occur in the tropics. During the wet season, colonies reduce brood rearing and abscond more. As the dry season begins, Africanized colonies expand rapidly and exhibit considerably higher rates of swarming, collect more pollen than nectar and become more defensive.

In Brazil, Panama and Costa Rica, a seasonal cycle is evident in swarming. Beekeepers have a seasonal cycle of honey production which Professor Ramirez in Costa Rica says can "produce abundant honey…if the colonies are properly managed." Correct management includes swarm control and feeding of dry sugar during the rainy season.

Gard Otis measured some of the seasonal differences in French Guiana. During the dry season, swarm-to-swarm intervals were short (mean 48.8 days) and afterswarms were common (mean 2.3 per colony). In the wet season, the swarm-to-swarm cycle was longer (64.5 days) with a lower number of afterswarms (1.3 mean). Absconding was more common in the wet season then dry.

One big difference in African tropical areas compared to American is that the seasons are much less predictable. Timing and management is a major difficulty for the beekeeper. The American tropics are more cyclic than those of Africa as they show marked wet and dry seasonal cycles and "have a predictability reminiscent of temperate areas" in the words of USDA scientists Tom Rinderer and Anita Collins.

Bee flight in tropical areas is more likely to be restricted by high temperature and heavy rain. Heat and excessive moisture and humidity may prevent plants from flowering or dilute nectar and pollen to the point that the plant is not worth foraging. High temperature may desiccate flowers or they may be open but contain no nectar or pollen. Bees often reduce flight at mid-day because of heat, humidity and lack of nectar or pollen—in the tropics this is "siesta time"—a sensible choice for bees and humans, given the intensity of the noontime sun.

In Venezuela, and in the Pacific areas of Central America from Panama to Mexico, the nectar flow period begins in November, almost opposite the day-length stimulation important for European bees. Close to the equator, there is little day length change throughout the year, perhaps so small as to make it of little value as a stimulatory influence. European bees carried from the U.S. to Venezuela for comparative studies did so poorly in the rainy season of April to August that the USDA lab "used exclusively Africanized bees…during this period" according to the researchers.

Brood rearing by bees in the tropics is continuous during the year, though it is reduced during the rainy season. Bee life during the rainy period is considerably shorter, 12–18 days; it extends to 20–25 days during the dearth period. Queens are more active and they live shorter lives. Drones are reared earlier, in greater abundance and are present longer during the year.

Tropical bees excel in the tropics. Africanized colonies have a probability of survival four times greater than European bees; only one Italian colony survived for 12 months without management, whereas 5 of 10 Africanized bee colonies were still going after 12 months without attention, according to a 1972 study by Professor Kerr and associates.

Maintaining Productive Colonies

Beekeepers are well aware that it takes populous colonies to produce honey crops. Africanized honey bees generally are highly responsive to environmental conditions. This means it is sometimes a challenge to manage Africanized bees as a fixed resource as practiced with temperate European bees. Africanized bees abscond, swarm and have a higher incidence of queen rearing.

Absconding

Absconding is abandonment of a nest by honey bees. When bees abscond, they form and move like a swarm. The entire population of bees leaves the nest when a colony absconds—they do not leave developing queen cells behind as is the case in swarming. In French Guiana, absconding occurred in 30% of colonies; a 24% absconding rate was reported for Costa Rica. In Africa, absconding rates are typically 15–30%, though up to 100% has been reported.

Two types of absconding behavior have been identified in response to two very different types of stimuli:

(1) **Disturbance-induced absconding** occurs as a response to such disturbances like partial or total nest destruction by predators such as ants and mammals, wax moth pressure, heavy wasp predation, fire in the area of the nest, rain on the nest or even beekeeper manipulations. Since this type of absconding usually occurs within a few hours after the disturbance it might be considered an alternative to stinging defense.

(2) **Resource-induced absconding** is a behavior the bees prepare for, much like swarming. It is prompted by lack of pollen/nectar/water resources. Virtually all the resources of the nest are used so only drawn comb is left behind by the bees. The workers eat all the eggs laid by the queen and rearing larvae stops prior to absconding. Virtually all the bees, except for a few stragglers, leave.

Absconding bees leave the nest site with nearly two times the amount of honey in honey stomachs compared to bees leaving in a swarm. The ages of bees in an abscond are more diverse and there are more older-aged bees in an abscond than in a swarm.

Why some bee colonies leave during resource-poor conditions and others stay is unknown. In Northeast Brazil a recent study reported at the Second International Conference on Africanized Honey Bees and Bee Mites reported that only about 10% of colonies do not abscond in that area. Ant attacks plus shortage of nectar or water seem to trigger absconding. Interestingly Mark Winston reported no differences in the amounts of honey, pollen, or brood/adult populations between colonies that abscond and those that do not. We have the same contradiction in swarming of temperate-zone bees—not all colonies swarm, though they may begin preparations to do so.

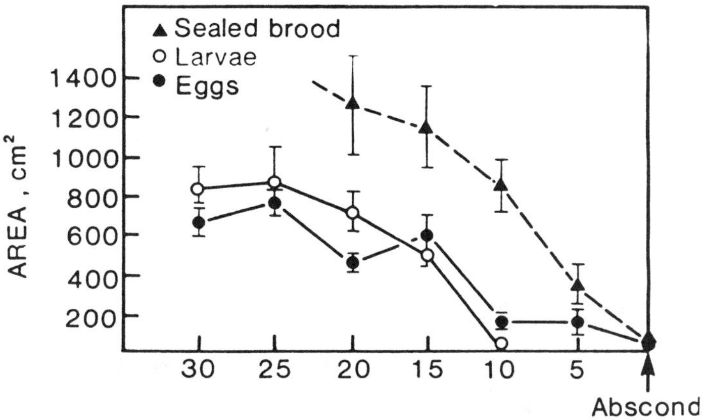

Figure 12–1. Areas of eggs, larvae and sealed brood in colonies of Africanized bees prior to absconding.
(From: Winston 1987. Biology of the Honey Bee)

Marla Spivak found absconding in Africanized but not European bees in her studies in Costa Rica. Most of the absconds were weak colonies that left during the resource-poor rainy season. In two years, she had 28 prime swarms issue from Africanized colonies and 8 absconds, nearly 60% of the total number of colonies in her study. During the period of Africanized bee absconding, European-race bees died due to dwindling.

A major difference between the African bee habitat in Africa and where the Africanized bee exists in South America. In Africa, the dearth period (when absconding occurs) is the dry season. When it is dry in one area, rain may occur in another region. Thus, it is adaptive behavior to leave (abscond) one area to seek another site because such an area exists.

In South America, the dearth period is the rainy season, which is a widespread phenomena. Leaving (absconding) will not necessarily take a migrating colony from a resource-poor area to another area with better nectar/pollen plants and less rain. Absconding in European or North American temperate bees also is not adaptive since climate and nectar/pollen resources are very similar over a wide area. Selective forces would not favor absconding, though perhaps some fall swarms may actually be absconds in North America.

Bees are not thinking creatures so they do not "know" dry versus rainy are two different situations. They simply respond to the stimulus of low resource availability. Leaving (absconding) was profitable to enough colonies in Africa over the 2000 or so years the eastern Africa climate has been similar to the one we see today (the age of the Sahara desert) so it was naturally selected. Bee colonies that absconded survived to pass on this behavior, through genetic inheritance, to the next generation. It was not adaptive in Europe, so temperate-zone bees did not pass the behavior along to subsequent generations.

Swarming

European bees in the tropics exhibited very little swarming. In contrast, swarming rates of Africanized bees are high. A single tropical bee colony may produce between 6 and 12 swarms and each colony additionally produces afterswarms. Gard Otis calculated a 16-fold growth rate of Africanized bees in French Guiana. For temperature-zone bees, one to 3.6 swarms per year is average.

Figure 12–2. Warily looking at small Africanized bee swarm. Beekeepers no longer attempt capture of such swarms.

Prime swarms in temperate-zone bees are 2.5 times larger than afterswarms. Gard Otis measured and weighed approximately equal numbers of Africanized bee swarms and afterswarms. Prime swarms averaged 1.57 kg (3.5 lb) and contained 16,030 bees; afterswarms 0.65 kg (1.4 lb) and 6910 bees. Prime swarms had more younger-aged bees, which carry more honey reserves, and fewer drones than afterswarms.

During favorable weather, colonies in French Guiana had short intervals between swarming episodes (mean of 48.8 days) and higher numbers of afterswarms (2.3) than during less favorable conditions (mean of 64.5 days and only 1.3 afterswarms). During the rainy season, only two colonies swarmed but a 30% absconding rate was found.

Otis conducted his studies using nuc boxes during Africanized bee colonization of French Guiana. Marla Spivak found a lower swarming rate in Costa Rica using larger hive boxes. In her study, Africanized bees swarmed three times more frequently compared to European-race colonies in the same apiary. Differences were especially noticeable at the lowest study elevation (900 m – 2950 ft) when 77% of Africanized bees but only 38% of European bees swarmed.

Swarms group together (megaswarms) in Africa at the beginning of the migratory phase. Such swarms may contain 100,000 workers and disperse great distances. On the wet Atlantic side of Panama, 22 megaswarms were seen, whereas on the Pacific side a higher number of smaller swarms were sighted, but no megaswarms. Does this signal migratory behavior of Africanized bees in the Americas? It is interesting to note that swarms on the Atlantic side of Mexico also are larger and fewer in number than those on the Pacific side.

Africanized bee colonies do not necessarily have ample stores of honey and pollen prior to swarming and they swarm at a different part of the seasonal cycle compared to temperate-zone bees. In Panama and Costa Rica, the vast majority of swarms emerge during favorable forage conditions beginning early in the dry season of December–February. On the wetter Atlantic side of Panama, the swarming pattern is bimodal. Peak swarming occurs in December-February and then again in May–June. Some swarms are megaswarms with larger than normal numbers of workers and queens.

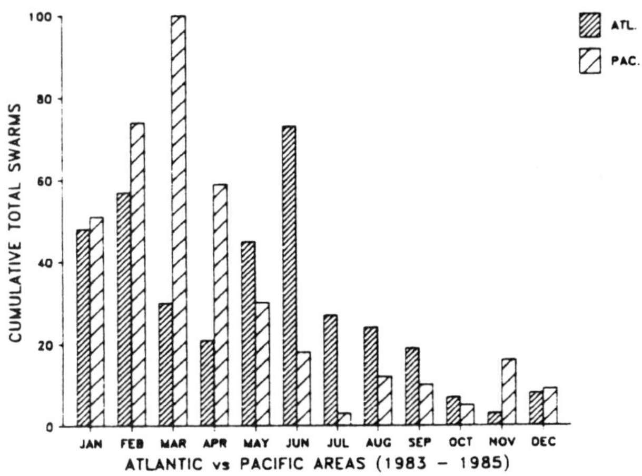

Figure 12–3. Swarming seasonality in Panama 1983–1985. (From: Boreham & Roubik 1987. Bull. Entomol. Soc. Amer. 33:34*)*

Large amounts of swarming and high absconding rates mean that the Africanized bee builds large numbers of feral colonies in a short time. They soon overwhelm managed colonies of European bees in their production of large numbers of drones that become available for mating with virgin queens. They are successful!

Colonizing

The ability of Africanized honey bees to colonize an area is well documented. Density of Africanized bee feral nests as high as 108/km^2 (280/mi^2) in open savanna of Brazil to between 10–20/km^2 (26–52/mi^2) in areas of moderately good foraging conditions in Venezuela and Peru have been reported. From 4 to 9 colonies/km^2 (10–23/mi^2) were found in three different areas in a southern Mexico survey 18 months after Africanized bee arrival.

Al Dietz, comparing studies conducted in Argentina and Mexico, concludes that "the pioneer populations of Africanized honey bees in Mexico displayed different behavioral patterns than the honey bees in Argentina." Pioneer colonies are "a different ecotype" that may respond by "absconding instead of increasing the level of defensive response."

Francis Ratnieks points out that colonizing traits "are unlikely to be well-suited for beekeeping." The reason hybridization may not occur when the Africanized bee moves into an area may be that the Africanized bee is well adapted to colonize whereas the European bee is not. According to Ratnieks, "beekeepers must make a sudden transition from the use of stocks that for up to about 100 generations (i.e., since the introduction of the movable frame hive) have been selected for good beekeeping traits to a stock that for about the same number of generations has been selected for bad beekeeping traits." (His assumption is that approximately 100 generations of Africanized bees have occurred in the Americas at the colonizing front.)

Bee colonies in the tropics have a different reproductive schedule. Honey storage is less critical; once resources improve, the stimulus is to divide via swarming rather than simply hoard food reserves. The bees migrate in year-round temperatures that permit flight, rather than store large surpluses in preparation for a cold, flightless winter season.

Tropical bees have evolved the evolutionary strategy of using resources to divide and reproduce. Large honey reserves make tropical bees a tempting target for predators. Enemies will get some of the colonies but some will survive to reproduce. In a familiar analogy—bee colonies do not "put all their eggs in one basket"!

If conditions are not favorable, the tropical bee tends to leave an area (abscond) and search for a new, more favorable place to live. The colony consumes all the pollen stores, stops rearing young (using the reduced number of eggs still being laid by the queen as food), and transfers honey stores to honey stomachs to depart enmass. The absconding swarm will find food somewhere else, or at least enough will, to naturally select for this behavior.

A second type of absconding also is part of tropical bee behavior. If the colony is disturbed, such as by raiding army ants, by fire, heavy smoking by the beekeeper, or a host of other disturbances, the tropical bee colony flees. They do not possess heavy honey stores, and evolutionary strategy for the behavior to stay and defend is not advantageous.

If small in number, the absconding bees can invade another colony (colony ursurpation) or, better yet, join other swarms in transit (megaswarm) and search for a new homesite. If a suitable cavity is not found, they settle under a large branch and build an exposed comb. Disturbance-induced absconding means leaving beeswax comb, brood, food resources and a secure nest site. Resource-induced absconding is a planned exit designed to leave few resources, (mainly empty comb) behind.

Defense

Hive defense is a prominent feature of tropical *scutellata* bees. They represent an extreme among bees in this behavior. The bees tend to exit the hive and defend outside the nest. This evolutionary strategy is advantageous because it allows small colonies to become large enough to utilize resources and reproduce.

It has been postulated that Africanized bee defensive behavior is highly developed because the native race of Africa has many predators that prey on honey bees. There are ants everywhere and wasps and hornets of several kinds. There are amphibians, birds and many types of mammals that seek and destroy bee nests. There is even the team approach of honey guide birds (*Indicator* sp.) leading mammals like baboons and honey badgers to ground or aerial nest locations, waiting until the mammal destroys the nest to take their share.

The greatest predators of all are humans and primates. Honey hunters of Africa are highly skilled in finding and then harvesting wild nests. Half the beeswax in the world comes from the part of the world where defensive bees live naturally. Honey and bee brood are consumed by many cultures, from the diminutive Pygmies to the very tall Masai. (Figure 3–3 page 32 is meant to depict Masai harvest of a feral bee colony.) Advanced cultures who keep bees in primitive domiciles still today destruct-harvest to utilize everything of the nest. They drive the bees away with fire and smoke or kill them first before exposing the entire nest. Honey hunters and beekeepers represent strong selective pressures on social bees.

Once acquired, characteristics such as defensiveness, reproductive potential and absconding behavior do not change rapidly without contrary selection pressure. Temperate bees have been kept by man intensely for maybe 500 years, but only in the last 150 years in a hive that permits convenient inspection and modification for economic betterment.

American beekeepers will probably change the Africanized bee. Change will not occur overnight. The most defensive colonies will be eliminated and traits selected that foster manageability. This has already happened in Brazil. Change comes more rapidly and becomes more widespread where there are government programs and incentives.

Do we manage a bee vastly different from the feral colony living in a hollow tree somewhere? The swarm we lose can still find and colonize a cavity and survive the winter, though only one in four will survive, much lower numbers than if we hive the swarm and place it in our apiary. Honey bees can survive without our intervention.

Is It a Subtropical Bee?

The subtropical regions of the world (China, Australia, Mexico and Argentina) are currently the major important honey-producing areas. Three major reasons for this are:

(1) temperate, manageable *Apis mellifera* bees can be used
(2) annual cycle is sufficiently predictable to permit management
(3) temperatures are high enough to permit year-round foraging and brood rearing.

In subtropical areas, daylength clues are available to key behaviors. There is annual variation since these regions are farther away from the equator. Nectar flows occur when daylength is increasing, though they are earlier in the cycle than for temperate areas. In Brazil, for example, the major flow of July to September is shortly after the shortest day in June.

The warmer year-round temperatures allow for more flight and less need for large quantities of honey stores for winter survival. There is enough regularity, especially where there is some agriculture (due to planting date and irrigation), to permit a successful seasonal management approach. Italian bees are excellent honey producers in the subtropics!

The Africanized honey bee arrived on the scene of important subtropical beekeeping areas only in the last few years. In Argentina, Africanized bee colonization has ceased and a transition zone has developed that has proven to be remarkably stable for several years. Will the northern distribution limit of Africanized bees in the U.S. occur at an equivalent latitude? For Texas that means between Houston and Dallas—about where they now are found.

USDA scientists Tom Rinderer and Rick Hellmich feel that rainfall is an important determining climate feature in the distribution of the African bee in Africa. Will southern and northern subtropical distribution of the Africanized bee be delimited by rainfall rather than temperature?

One factor that may be important is the amount of, or lack of, gene flow between bees. In Central America, the Africanized bee encountered virtually no feral population of European bees and less than an estimated 500,000 managed colonies (by Orley Taylor's calculations) so that "the majority (>95%) of virgin queens from feral African[ized] colonies could only have mated with feral (i.e. African) drones."

Could the subtropical regions, with higher numbers of feral colonies, a less tropical climate, more beekeepers with European bees, and bees that are better adapted (at least from the beekeeper's perspective) form a barrier that will limit Africanized bee dominance?

We have learned much about Africanized bees but our ignorance is still considerable!

The Ultimate Myth—Feral Bees?

We have replaced one bee (German) with another (Italian) in the U.S. The last myth listed in Chapter 5 was that the U.S. has a large feral population of honey bees occupying the best nesting sites. If advance of the Africanized bee is to be halted, we need to concentrate on a bee that might best be able to compete with Africanized bees on their own terms.

A study by Schiff demonstrated that the feral bee population in the U.S. is different from beekeeper-managed bees. The study documented that beekeepers maintain colonies that are largely Italian and Carniolan in genetic composition; 22% of feral colonies vary more in their genetic makeup with a number of genetic markers indicating German ancestry.

Mark Winston presented an interesting perspective to the issue in a 1995 *Bee Culture* column.

> "The death of many, if not most, feral colonies because of mites, coupled with reduced numbers of managed colonies due to mite damage and economic factors, could provide a real opportunity for Africanized bees to become well established in the wild...[beekeeper selection] continues to push managed honey bees toward large, docile, relatively non-swarming traits.....Natural selection pushes feral colonies to be smaller, swarm more frequently and be more aggressive in order to survive in the wild."

Dave Fletcher believes we need to look to adding more genetic variation to the population of Africanized bees in the Americas. Francis Ratnieks believes that, at the limit of its range, it might be easier to bring European genes into the colonizing population. The irony of this, as Ratnieks points out, is that the bee that colonizes is not the best bee for beekeepers to manage. One hundred generations of selection "for good beekeeping traits" has occurred in European bees "but the same number of generations has been selected for bad beekeeping traits" by natural selection in Africanized bees.

After 30 years we have not developed a better Africanized bee for beekeepers, except maybe in Brazil.

A New Resource

We have come to see that tropical bees are less manageable, less tolerant of examination, and less amenable to being moved. To learn how better to manage bees in the tropics requires that we change and adapt to better accommodate to the bees' needs. This has been the formula for success for temperate area beekeepers—why should it be different for tropical bees?

Honey bees, whether temperate-zone or tropical, benefit from the changes humans bring. We cut down trees and open up spaces for agriculture and homes by changing the native vegetation. Honey bees do well in the edge between open areas and forest. They have fewer competitors for more cavities in which to build nests. They also benefit from many of the introduced plants by gaining concentrated foraging areas.

In the tropics, honey bees become a new resource for human economic gain. Honey hunting is feasible since there now is a large feral population of honey bees. There were few before the Africanized bee because European bees did not swarm in the tropical climate. Also swarms are now present year-round. Harvest potential from the introduced *Apis mellifera* is 100 times greater than with the native stingless honey bee.

Figure 12–4. José Cus (Belize) with harvest from feral Africanized bee colony.

Traditional beekeepers find swarms of little value since they cannot be managed in the sense that swarms of European colonies can be, but new beekeepers hive some, with some failures, and thus a new population of beekeepers develops. Local people do not let a valuable resource go to waste in the tropics, as we do not in our own neighborhoods.

Africanized Bees—A Success Story?

Is the U.S. beekeeper going to be able to maintain a population of honey bees, distinct from the feral population? Should this population be a temperate-zone bee, a changed (less defensive, reduced swarming, less absconding, more productive) tropical bee or a mixture of both?

Do we wish to attempt to introduce bees from one area to another again? We might be able to develop a bee better suited to the subtropics by crossbreeding but can we do it for temperate areas? There is variation in the population of tropical bees but the greater diversity is in Africa not in the Americas. Are the risks of exotic introductions less now that mites have been carried worldwide? Are there other pests and diseases that we might yet introduce —never again to exterminate?

On the other hand perhaps the key to more rapid development of resistant populations to mites and less defensive bees is in such an approach. Do we have the moral right to threaten the existence of native species by introducing an organism from one area to another? Most introductions do not seem to be successful, yet the next introduction may be the solution, similar to Italian bee success in the mid-1800s.

Biocontrol systems have a carefully thought-out protocol connected with an introduction of a new organism. Because the honey bee is considered a beneficial species does it somehow permit it to live beyond the rules? Importers of bumble bees for immensely profitable greenhouse pollination think bumble bees should warrant relaxed rules to bring more species to the U.S. Should honey bee rules be less stringent?

It is not correct to assume the Africanized bee is without value. Campesinos and beekeepers in Central and South America have found many good characteristics:

- They are better pollinators.
- They are better honey producers—under some circumstances.
- The Africanized bee is more resistant to Varroa mite than European bees.
- Funding to learn about Africanized bees has helped teach us about bees in general.
- African bee stock is less defensive in cooler climates.

What lessons can we learn from African bee importation into South America? The bee is not all bad but neither is it necessarily good. Brazilian beekeepers may prefer the Africanized bee but then they really no longer have a choice.

A Different Bee

Tropical bees are different. It is evident that the American tropics are climatically close enough to the area in Africa where this bee evolved to permit the Africanized bee to prosper. They have proceeded to colonize virtually all of the Americas because they proved to be the better bee ecotype than the earlier-introduced European bees.

Figure 12–5. Beekeeping Africanized bees—possible but different from European culture.

Africanized bees are better adapted to successfully exploit unpredictable and variable-resource conditions. The extent to which European bees were unsuccessful in the Americas and in many introductions into Africa reflects the limited capacity of European bees to adjust to the conditions encountered when carried to tropical lands.

Tropical bees have been shaped by different selection factors. They live in a different environment. Some of the same factors continue to shape and mold the populations of the Africanized bee. At the frontiers of colonization, the pioneer swarms are different, not at all like the bees that exist in the areas invaded previously. There may be some mixing of the new and the old but selection pressure favors those biologies that promote colonization. Once established, the better-adapted ecotype succeeds. Unfortunately from our perspective, the better bee is the one more similar to the parent stock from Africa.

The Africanization process leads not to a hybrid but to the African bee. It is the better bee physically, biologically and behaviorly. It is a bee better adapted for living in the tropics. Though it is not a hybrid, the Africanized bee is slightly different, not pure African and it will probably change over time.

The Africanized bee is not a myth. If nothing else, the Africanized bee is a success story of unprecedented magnitude.

✧✧✧✧✧✧✧✧✧

And so I conclude this discourse on Africanized bees. I hope my experiences and the experiences of others has been enlightening and informative. The Africanized bee is a most unique phenomenon. We have not heard the last of it. It has caused some problems for beekeepers, some real tragedies for families of victims and economic hardship for beekeepers and others whose lives have been touched by this population of bees. It will not go away. We need to learn to better live with it. This book is dedicated to that goal.

Good beekeeping! May all your encounters with Africanized bees, and bees in general, be pleasurable and positive.

REFERENCES

References

Chapter 1

De Jong, D. 1994. The African experience. *Bee Culture* 122(8):453, 456–461.

Dietz, A. 1992. Honey bees of the world. In: *The Hive and the Honey Bee*. Dadant & Sons, Inc., Hamilton, IL. Ch. 2. 23–71.

Flakus, G. 1993. *Living* with *Killer Bees*. Quick Trading Co., Oakland, CA. 144 pp. (Available from A. I. Root Co.)

Gore, R. 1976. Those fiery Brazilian bees. *National Geographic* 149(4):490–501.

Gonçalves, L. S., A. C. Stort and D. De Jong. 1991. Beekeeping in Brazil. In: *The "African" Honey Bee.* Ed. M. Spivak, D. J. C. Fletcher and M. D. Breed. Westview Press, Boulder, CO. Ch.18. 359–372.

Kerr, W. E. 1957. Introdução de abelhas africanas no Brasil. *Brasil Apicola* 3:211–213.

Kerr, W. E. 1967. The history of the introduction of Africanized bees into Brazil. *So. Afr. Bee Jour.* 39:3–5.

Kerr, W. E., L. C. Gonçalves, A. C. Stort and D. Bueno. 1967. Biological and genetical information on *Apis mellifera adansonii. Proc. XXI International Beekeeping Congress* (Maryland). 495–496.

McGregor, S. E. 1970. The African bee and American beekeeping. *Amer. Bee Jour.* 110:460–461.

Michener, C. 1975. The Brazilian bee problem. *Ann. Rev. Entomol.* 20:339–416.

Morse, R. A. 1994. The African experience. *Bee Culture* 122(8):453, 456–460.

National Academy of Sciences (NAS). 1972. *Final Report. Committee on the African Honey Bee.* PB-211. 952. National Research Council, National Academy of Sciences. Tech. Info. Service, U. S. Dept. of Commerce, Springfield, VA. 95 pp.

Rinderer, T. E. 1988. Evolutionary aspects of the Africanization of honey bee populations in the Americas. In: *Africanized Honey Bees and Bee Mites*, Ed. G. Needham, R. E. Page, Jr., M. Delfinado-Baker and C. Bowman. John Wiley & Sons, Inc., NY. 13–28.

Sanford, M. T. 1995. *Information on the African honey bee.* UFL ENY–114 Fact Sheet. 6 pp.

Seegers, S. and K. Seegers. 1976. The truth about those "killer bees." *Reader's Digest* 108(645):25–28, 31.

Schiff, N. M., W. S. Sheppard, G. M. Loper and H. Shimanuki. 1994. Genetic diversity of feral honey bee (Hymenoptera: Apidae) populations in the southern United States. *Ann. Ento. Soc. Amer.* 87(6):842–848.

von Frisch, K. 1967. *The Dance Language and Orientation of Bees.* Belknap Press (Harvard Univ. Press), Cambridge, MA. 566 pp.

Winston, M. L. 1992. *Killer Bees.* Harvard Univ. Press, Cambridge, MA. 162 pp.

Chapter 2

Allen, M. 1996. Global report. *Bee Biz* 1(3):35.

Anonymous. 1999. AHB headed to central valley? *Calif. Agric.* 53(2):4.

Burgett, M., S. Shorey, J. Cordara, G. Gardiol and W. S. Sheppard. 1995. The present status of Africanized honey bees in Uruguay. *Amer. Bee Jour.* 135(5):328–330.

Cox, R. 1994. USDA Interagency meeting on Africanized honey bees. *Amer. Bee Jour.* 134(10):666–668.

Dietz, A., R. Kreil and J. Pettis. 1986. The potential limit of survival for Africanzied bees in the United States. *Proc. Africanized Honey Bee Symp.* Amer. Farm Bureau. Atlanta, GA. 87–100.

Eischen, F. A. 1994. How far north will Africanized bees go? *Amer. Bee Jour.* 134(8):522–523.

Eischen, F. A. 1994. Texas and Africanized bees: dealing with the press. *Amer. Bee Jour.* 134(12):801–802. (See also related articles in Feb., April, June issues of *Amer. Bee Jour.*)

Eischen, F. A. 1999. Varroa and Africanized bees: an update. *Amer. Bee Jour.* 139:594–595.

Flakus, G. 1993. *Living* with *Killer Bees.* Quick Trading Co., Oakland, CA. 144 pp. (Available from A. I. Root Co.)

Fletcher, D. J. C. 1978. The African bee, *Apis mellifera adansonii*, in Africa. *Ann. Rev. Entomol.* 23:151–171.

Fletcher, D. J. C. 1991. Interdependence of genetics and ecology in a solution to the African bee problem. In: *The "African" Honey Bee*, Ed. M. Spivak, D. J. C. Fletcher and M. D. Breed. Westview Press, Boulder, CO. Ch. 4. 77–94.

Flottum, K. 1999. Apimondia '99—Gleanings from the experts—African honey bees. *Bee Culture* 127(11):26–28.

Gary, N. S., H. V. Daly, S. Lock and M. Race. 1985. The Africanized honey bee: ahead of schedule. *Calif. Agric.* 39:4–7.

Guzman-Novoa, E. and R. E. Page, Jr. 1994. The impact of Africanized bees on Mexican beekeeping. *Amer. Bee Jour.* 134(2):101–106.

Guzman-Novoa, E. and R. E. Page, Jr. 1999. Selective breeding of honey bees (Hymenoptera: Apidae) in Africanized areas. *Jour. Econ. Entomol.* 92(3):521–525.

Jackson, P. 1994. African honey bees still on a roll. *Bee Culture* 122(11):620–621.

Kaplan, K. J. 1990. The Africanized honey bees. *Agric. Research* USDA, ARS. 38(12):4–11.

Kempff Mercado, N. 1971. Contribucion a la solucion del problema de las abejas africanas. *Gac. Colmenar* 33:87–92.

Kent, R. 1989. The African honey bee in Peru: an insect invader and its impact on beekeeping. *Appl. Geog.* 9:237–257. (Reprinted with minor changes as "The Africanized bee in Peru." In: *The "African" Honey Bee*, Ed. M. Spivak, D. J. C. Fletcher and M. D. Breed. Westview Press, Boulder, CO. Ch. 18. 334–346.

Kerr, W. E. 1957. Introdução de abelhas africanas no Brasil. *Brasil Apicola* 3:211–213.

Kerr, W. E. 1967. The history of the introduction of Africanized bees into Brazil. *So. Afr. Bee Jour.* 39:3–5.

Kerr, W. E., L. C. Gonçalves, C. Stort and D. Bueno. 1967. Biological and genetical information on *Apis mellifera adansonii. Proc. XXI International Beekeeping Congress* (Maryland). 495–496.

Kerr, W. E., S. L. del Rio, M. D. Barrionuevo. 1982. The southern limits of distribution of the Africanized honey bee in South America. *Amer. Bee Jour.* 122:196–198.

Kraus, B. and R. E. Page, Jr. 1995. Effect of *Varroa jacobsoni* (Mesostigmata: Varroidae) on feral *Apis mellifera* (Hymenoptera: Apidae). *California Environ. Entomol.* 24(6):1473–1480.

Laidlaw, H. H. 1977. The importation of semen from Brazil into California. *Amer. Bee Jour.* 117(3):153.

Langston, D. 1996. Africanized honey bees in Arizona. *Training Manual* AZ Coop. Extension. (Includes VHS and slides).

Lobo, J. A., M. A. del Lama and M. A. Mestriner 1989. Population differentiation and racial admixture in the Africanized honey bee (*Apis mellifera* L.). *Evolution* 43:794–802.

Loper, G. M. 1997. Genetic evidence of the Africanization of feral colonies in southern Arizona between 1993–1995. *Amer. Bee Jour.* 137:669–671.

Mangum, W. A. 2000. Adventures in Bolivian beekeeping. *Amer. Bee Jour.* 140(10):767–770.

Moffett, J. O., D. L. Maki, T. Andre and M. M. Fierro. 1987. The African-Africanized bee in Chiapas, Mexico. *Amer. Bee Jour.* 127:517–519, 525.

Marston, G. 1985. A beekeeper examines the Africanized bee quarantine in California and how it has affected his beekeeping. *Amer. Bee Jour.* 125(12):833–834.

Morse, R. A. 1994. *The New Complete Guide to Beekeeping.* Countryman Press, Inc., Woodstock, VT. 207 pp.

Morse, R. A., D. M. Burgett, J. T. Ambrose, W. E. Conner, R. D. Fell. 1973. Early introductions of African bees in Europe and the New World. *Bee World* 54:57–60.

Moffett, J. 1995. African bee hampers pollination in Texas. *Bee Biz* 1(1):8–9.

Nogueira-Neto, P. 1964. The spread of the fierce African bee in Brazil. *Bee World* 45:119–121.

ORISA (Organismo Internat. Regional de Sanidad Agropecuaria). 1988. *Manejo y control de la abeja Africanizada.* ORISA, San Salvador, El Salvador. 229 pp.

Page, R. E. 1992. How Africanized honey bees will affect California agriculture. *Calif. Agric.* 46(1):18–19.

Page, R. E. 1995. Early stages of Africanized honey bee "invasion" detected in Southern California. *Calif. Almond News* Aug. 95:2.

Page, R. E. 1998. Blessing or curse? Varroa mite impacts Africanized bee spread and beekeeping. *Calif. Agric.* 52(2):7–12.

Rinderer, T. E. and R. L. Hellmich, II. 1991. The process of Africanization. In: *The "African" Honey Bee.* Ed. M. Spivak, D. J. C. Fletcher and M. D. Breed. Westview Press, Boulder, CO. Ch. 5. 95–118.

Rinderer, T. E. and H. Shimanuki. 1993. Africanized honey bees in the USA: where do we go from here? *Bee World* 74:157–159.

Rodriguez-Lainz, A., C. L. Feitz and V. Kramer. 1998. Africanized honey bees in California: public health aspects. *Calif. Morbidity Report* Nov. 1998. 4–7.

Rubink, W. L. 1999. Varroa mites and the population biology of Africanized honey bees in Mexico and Texas. *Proc. 36th Inter. Apimondia Congress.* Vancouver, Canada. 206.

Rubink, W. L., E. A. Sugden, P. Luevano-Martinez, A. M. Collins and W. T. Wilson. 1994. Subtropical honey bee swarming dynamics before and after Africanization in northeastern Mexico and southern Texas. *Proc. 5th International Conf. Apic. Trop. Clim.* IBRA, Cardiff, Wales. 186–192.

Rubink, W. L., W. T. Wilson, A. M. Collins, and P. Luevano-Martinez. 1995. Comparative Africanization rates in feral honey bee populations at three latitudes in northeastern Mexico and southern Texas. *Amer. Bee Jour.* 135(12):829.

Sanford, M. T. 1999. African honey bee surprises again. *Apis* 17(5):1–4.

Sanford, M. T. 2000. The African honey bee in Ecuador: can paradise be maintained? *Apis* 18(9):1–3.

Schmidt, J. O. 1995. Biological control of Africanized honey bees with swarm traps: economics and logistics. *Amer. Bee Jour.* 135(12): 829–830.

Sheppard, W .S., A. E. E. Soares, D. De Jong, and H. Shimanuki. 1991. Hybrid status of honey bee populations near the historic origin of Africanization in Brazil. *Apidologie* 22:643–652.

Southwick, E. E., D W. Roubik and J. M. Williams 1990. Comparative energy balance in groups of Africanized and European honey bees: ecological implications. *Comparative Biochem. Physiol.* 97A:1–7.

Sugden, E. A., A. M. Collins, W. L. Ruink, L. A. Bradley, K. Davis and E. Quintero. 1994. Two years of "Africanization"; honey bee stinging incidents in Texas. *BeeScience 3* (2):68–75.

Sylvester, H. A., T. E. Rinderer and H. Shimanuki. 1992. Certification options for dealing with Africanized bees. *Amer. Bee Jour.* 132(3):182–185.

Taber, S. 1977. The African bee in Louisiana. *Amer. Bee Jour.* 117(3):152–154.

Taylor, O. R. 1977. The past and possible future spread of Africanized honeybees in the Americas. *Bee World* 58:19–30.

Taylor, O. R. 1985. African bees: potential impact in the United States. *Bull. Entomol. Soc. Amer.* 31:15–24.

Taylor, O. R. 1999. Displacement of European honey bee subspecies by an invading African subspecies in the Americas. *Proc. Apic. 21st Century* (Michigan State Univ.). Wicwas Press, Cheshire, CT. 38–46.

Taylor, O. R. and M. Spivak. 1984. Climatic limits of tropical African honey bees in the Americas. *Bee World* 65:15–24.

Vaissiere, B. E. and F. G. Maxwell 1989. Potential impact of the Africanized honey bees in Texas. *Texas Afr. Honey Bee Program.* Mimeo 27 pp.

Visscher, P. K., R. S. Vetter and F. C. Baptista. 1995. The progress of Africanized bees in the United States (1990–1995). *Calif. Agric.* 51:22–25 (also web: http://bees.ucr.edu).

Winston, M. L. 1992. *Killer Bees.* Harvard Univ Press, Cambridge, MA. 162 pp.

Woyke, J. 1969. African honey bees in Brazil. *Amer. Bee Jour.* 109(9):342–344.

Chapter 3

Boreham, M. M. and D. W. Roubik. 1987. Population change and control of Africanized honey bees (Hymenoptera: Apidae) in the Panama Canal area. *Bull. Entomol. Soc. Amer.* 33:34–39.

Danka, R. G. and T. E. Rinderer. 1988. Social reproductive parasitism by Africanized honey bees. In: *Africanized Honey Bees and Bee Mites.* Ed. G. Needham, R. E. Page, Jr., M. Delfinado-Baker and C. Bowman. John Wiley & Sons, Inc., NY. 214–222.

DeGrandi-Hoffman, G. and S. S. Schneider. 2001. Worker behaviors in queenless Africanized honey bee colonies. *Proc. 2nd Internat. Conf. on Africanized Honey Bees and Bee Mites.* To be published by A. I. Root Co., Medina, OH.

DeGrandi-Hoffman, G., J. C. Watkins, A. M. Collins, G. M. Loper, J. H. Martin, M. C. Arias and W. S. Sheppard. 1998. Queen developmental time as a factor in the Africanization of European honey bee (Hymenoptera: Apidae) populations. *Ann. Ent. Soc. Amer.* 91:53–58.

Dietz, A. 1992. Honey bees of the world. In: *The Hive and the Honey Bee.* Dadant and Sons, Inc. Hamilton, IL. Ch. 2. 23–71.

Fewell, J. and J. Harrison. 2001. Differences in worker task phenotypes of African and European honey bees. *Proc. 2nd Internat. Conf. Africanized Honey Bees and Bee Mites.* To be published by A. I. Root Co., Medina, OH.

Flakus, Greg. 1993. *Living* with *Killer Bees.* Quick Trading Co., Oakland, CA. 144 pp. (Available from A. I. Root Co)

Fletcher, D. J. C. 1978. The African bee, *Apis mellifera adansonii,* in Africa. *Ann. Rev. Entomol.* 23:151–171.

Fletcher, D. J. C. 1991. Interdependence of genetics and ecology in a solution to the African bee problem. In: *The "African" Honey Bee.* Ed. M. Spivak, D. J. C. Fletcher and M. D. Breed. Westview Press, Boulder, CO. Ch. 4. 77–94.

Gonçalves, L. S., A. C. Stout and D. De Jong. 1991. Beekeeping in Brazil. In: *The "African" Honey Bee.* Ed. M. Spivak, D. J. C. Fletcher and M. D. Breed. Westview Press, Boulder, CO. Ch. 18. 334–346.

Guy, R. D. 1972. Commercial beekeeping with African bees. *Bee World* 53:14–22, 159–166.

Harrison, J. F. and H. G. Hall. 1993. African-European honeybee hybrids have low nonintermediate metabolic capacities. *Nature* 363:258–260.

Kempff Mercado, N. 1971. Contribucion a la solucion del problema de las abejas Africanas. *Gac. Colmenar* 33:87–92.

Kerr, W. E. 1957. Introduçae de abelhas africanas no Brasil. *Brasil Apicola* 3:211–213.

Kerr, W. E. 1967. The history of the introduction of Africanized bees into Brazil. *So. Afr. Bee Jour.* 39:3–5.

Kerr, W. E. and V. de Portugal-Araújo. 1958. Raças de abelhas de Africa. *Garcia de Orta* 6:53–59.

Loper, G. M., J. Fewell, D. R. Smith, W. S. Sheppard and H. Schiff. 1999. Genetic changes of a population of feral honey bees in the Sonoran desert of southern Arizona following the arrival of *Acarapis woodi, Varroa jacobsoni* and Africanization. *Proc. Apiculture for 21st Century.* (Michigan State University) Wicwas, Press, Cheshire, CT. 47–51.

Michener, C. D. 1975. The Brazilian bee problem. *Ann. Rev. Entomol.* 20:399–416.

Morse, R. A., D. M. Burgett, J. T. Ambrose, W. E. Conner, and R. D. Fell. 1973. Early introductions of African bees in Europe and the New World. *Bee World* 54:57–60.

Nogueira-Neto, P. 1964. The spread of the fierce African bee in Brazil. *Bee World* 45:119–121.

Otis, G. W. 1982. Population biology of the Africanized honey bee. *Proc. Sym. I.U.S.S.I. Social Insects in the Tropics.* Mexico. 209–219.

Otis, G. W. and S. Hadisoesilo. 1999. Insights into honey bee biology from *Apis nigrocincta* of Indonesia. *Proc. Apiculture for 21st Century.* (Michigan State University) Wicwas Press, Cheshire, CT. 69–79.

Portugal-Araújo, V. de. 1971. The central African bee in South America. *Bee World* 52:116–121.

Rinderer, T. E. and R. L. Hellmich, II. 1991. The process of Africanization. In: *The "African" Honey Bee.* Ed. M. Spivak, D. J. C. Fletcher and M. D. Breed. Westview Press, Boulder, CO. Ch. 5. 95–118.

Rinderer, T. E., R. L. Hellmich, R. G. Danka and A. M. Collins. 1985. Male repoductive parasitism: A factor in the Africanization of European honey-bee populations. *Science* 228:1119–1121.

Rinderer, T. E., J. A. Stelzer, B. P. Oldroyd, S. M. Buco and W. L. Rubink. 1991. Hybridization between European and Africanized honey bees in the neotropical Yucatan Peninsula. *Science* 253:309–311.

Rinderer, T. E., J. E. Wright, H. Shimanuki, F. Parker, E. Erickson and W. T. Wilson. 1987. The proposed honey-bee regulated zone in Mexico. *Amer. Bee Jour.* 127:160–164.

Roubik, David. 1991. Aspects of Africanized honey bee ecology in tropical America. In: *The "African" Honey Bee.* Ed. M. Spivak, D. J. C. Fletcher and M. D. Breed. Westview Press, Boulder, CO. Ch. 13. 259–282.

Ruttner, F. 1975. Races of bees. In: *The Hive and the Honey Bee.* Dadant and Sons, Inc., Hamilton, IL. Ch. 2.

Ruttner, F. 1987. *Biography and Taxonomy of Honeybees.* Springer-Verlag, Berlin. 284 pp.

Schiff, N. M. and W. S. Sheppard 1993. Mitochondrial DNA evidence for the 19th century introduction of African honey bees into the United States. *Experientia* 49:550–552.

Schnetler, E. A. 1946. Honey production record. *So. African Bee Jour.* 21:10.

Sheppard, W. S. 1989. A history of the introduction of honey bee races into the United States. *Amer. Bee Jour.* 129:664–667.

Smith, F. G. 1961. The races of honeybees in Africa. *Bee World* 42:255–260.

Spivak, M., D. J. C. Fletcher, & M. D. Breed. 1991. Introduction. In: *The "African" Honey Bee*. Ed. M. Spivak, D. J. C. Fletcher and M. D. Breed. Westview Press, Boulder, CO. Introduction. 1–9.

Taylor, O. R. 1985. African bees: potential impact in the United States. *Bull. Entomol. Soc. Amer.* 31:15–24.

Taylor, O. R. 1999. Displacement of European honey bee subspecies by an invading African subspecies in the Americas. *Proc. Apiculture for 21st Century*. (Michigan State University) Wicwas Press, Cheshire, CT. 38–46.

Vergara, C., A. Dietz and A. Perez. 1989. Usurpation of managed honey bee colonies by migratory swarms in Tabasco, Mexico. *Amer. Bee Jour.* 129:824–825.

Winston, M. L. 1992. *Killer Bees*. Harvard Univ. Press, Cambridge, MA. 162 pp.

Woyke, J. 1969. African honey bees in Brazil. *Amer. Bee Jour.* 109(9):342–344.

Chapter 4

Boreham, M. M. and D. W. Roubik. 1987. Population change and control of Africanized honey bees (Hymenoptera: Apidae) in the Panama Canal area. *Bull. Entomol. Soc. Amer.* 33:34–39.

Buco, S. M. T. E. Rinderer, H. A. Sylvester, A. M. Collins, V. A. Lawrence and R. M. Crewe. 1987. Morphometric differences between South American Africanized and South African (*Apis mellifera scutellata*) honey bees. *Apidologie* 18:217–222.

Carlson, D. A. 1988. Africanized and European honey-bee drones and comb waxes: Analysis of hydrocarbon components for identification. In: *Africanized Honey Bees and Bee Mites*. Ed. G. Needham, R. E. Page, Jr., M. Delfinado-Baker and C. Bowman. John Wiley & Sons, Inc., NY. 264–274.

Collins, A. M., H. V. Daly, T. E. Rinderer, J. R. Harbo, and K. Hoelmer. 1994. Correlations between morphology and colony defense in *Apis mellifera* L. *Jour. Apic. Res.* 33(1):3–10.

Daly, H. V. 1998. Overview of the identification of Africanized honey bees. In: *Africanized Honey Bees and Bee Mites*. Ed. G. R. Needham, R. E. Page, M. Delfinado-Baker, C. E. Bowman. Ellis Horwood Ltd., Chichester, England. 245–247.

Daly, H. V. 1991. Systematics and identification of Africanized honey bees. In: *The "African" Honey Bee*. Ed. M. Spivak, D. J. C. Fletcher and M. D. Breed. Westview Press, Boulder, CO. 13–44.

Daly, H. V. & S. S. Balling. 1978. Identification of Africanized honey bees in the Western hemisphere by discriminant analysis. *Jour. Kansas Entomol. Soc.* 51:857–869.

Daly, H. V. and R. A. Morse. 1991. Abnormal sizes of worker honey bees (*Apis mellifera* L.) reared from drone comb (Hymenoptera: Apidae). *Jour. Kansas Ento. Soc.* 64(2):193–196.

Daly, H. V., D. De Jong and N. D. Stone. 1988. Effect of parasitism by *Varroa jacobsoni* on morphometrics of Africanized worker honey bees. *Jour. Apic. Res.* 27(2):126–130.

Daly, H. V., R. C. Danka, K. H. Hoelmer, T. E. Rinderer and S. M. Buco. 1995. Honey bee morphometrics: linearity of variables with respect to body size and classification tested with European worker bees reared by varying ratios of nurse bees. *Jour. Apic. Res.* 34(3):129–145.

Dietz, A. 1992. Honey bees of the world. In: *The Hive and the Honey Bee.* Dadant and Sons, Inc., Hamilton, IL. Ch. 2. 23–71.

Fletcher, D. J. C. 1991. Interdependence of genetics and ecology in a solution to the African bee problem. In: *The "African" Honey Bee.* Ed. M. Spivak, D. J. C. Fletcher and M. D. Breed. Westview Press, Boulder, CO. Ch. 4. 77–94.

Guzman-Novoa, E., and R. E. Page. 1994. The impact of Africanized bees on Mexican beekeeping. *Amer. Bee Jour.* 134:101–106.

Guzman-Novoa, E. and R. E. Page. 1994. Genetic dominance and worker interactions affect honeybee colony defense. *Behav. Ecol.* 5(1):91–97.

Guzman-Novoa, E., R. E. Page and M. K. Fondrk. 1994. Morphometric techniques do not detect intermediate and low levels of Africanization in honey bee (*Apis mellifera* L.) colonies. *Ann. Entomol. Soc. Amer.* 87:507–515.

Hall, H. G. 1990. Parental analysis of introgressive hybridization between African and European honeybees using nuclear DNA RFLPs. *Genetics* 125:611–621.

Hall, H. G. 1991. Genetic characterization of honey bees through DNA analysis. In: *The "African" Honey Bee.* Ed. M. Spivak, D. J. C. Fletcher and M. D. Breed. Westview Press, Boulder, CO. Ch. 3. 45–76.

Hall, H. G. 1999. Genetic and physiological studies of African and European honey bee hybridization: Past, present, and into the 21st century. *Proc. Apiculture for the 21st Century.* (Michigan State University) Wicwas Press, Cheshire, CT. 52–59.

Hall, H. G. 1998. PCR amplification of a locus with RFLP alleles specific to African honey bees. *Biochem. Genetics* 36:351–361.

Hall, H. G. and K. Muralidharan 1989. Evidence from mitochondrial DNA that African honey bees spread as continuous maternal lineages. *Nature* 339:211–213.

Hall, H. G., and D. R. Smith. 1991. Distinguishing African and European honey bee matrilines with amplified mitochondrial DNA. *Proc. Natl. Acad. Sci.* 88:4874–4877.

Hall, H. G., A. Suazo and O. R. Taylor. 2000. African honey bee intro gression into southern Texas: Parental analyses using DNA markers. *Proc. 35th Internat. Apimondia Congress.* Vancouver, Canada. 204–205.

Harrison, J. F., and H. G. Hall. 1993. African-European honeybee hybrids have low non-intermediate metabolic capacities. *Nature* 363:258–260.

Herbert, E. W., H. A. Sylvester, J. D. Vanderberg, and H. Shimanuki. 1988. Influence of nutritional stress and the age of adults on the morphometrics of honey bees (*Apis mellifera* L.). *Apidologie* 19:221–229.

Houck, M. A. 1990. Bee morphometrics and the influence of size on the discrimination of Africanized and non-Africanized bees. *Amer. Bee Jour.* 130:802–803.

Kerr, W., and D. Bueno. 1970. Natural crossing between *Apis mellifera adansonii* and *Apis mellifera ligustica*. *Evolution* 24:145–148.

Levine, B. and D. Carlson. 1987. Species identification through chemical analysis. *Anal. Chem.* 56(6):468–471.

Lobo, J. A. & H. Krieger. 1992. Maximum likelihood estimates of gene frequencies and racial admixture in *Apis mellifera* L. (Africanized honey bees). *Heredity* 68:441–448.

Lobo, J. A., M. A. Del Lama and M. A. Mestriner, 1989. Population differentiation and racial admixture in the Africanized honey bee (*Apis mellifera* L). *Evolution* 43:794–802.

Nabors, R. A. and S. K. Bajaj. 1998. A measurement technique with potential to screen specimens of *Apis mellifera* L. for subsequent Africanization determination. *Amer. Bee Jour.* 138:56–57.

Nielsen, D. I., P. R. Ebert, G. J. Hunt, E. Guzman-Novoa, S. A. Kinnee, and R. E. Page, Jr. 1999. Identification of Africanized honey bees (Hymenoptera: Apidae) incorporating morphometrics and an improved polymerase chain nectar mittotyping procedure. *Ann. Entomol. Soc. Amer.* 92:167–174.

Page, R. E. 1989. Neotropical African bees. *Nature* 339:181–182.

Quezada-Euan, J. G, and D. M. Hinsull. 1995. Evidence of continued European morphometrics and mtDNA in feral colonies of honey bees (*Apis mellifera*) from the Yucatán peninsula, Mexico. *Jour. Apic. Res.* 34:161–166.

Quezada-Euan, J. G., and L. M. Medina. 1998. Hybridization between European and Africanized honey bees (*Apis mellifera* L) in tropical Yucatán. *Apidologie* 29:555–568.

Ratnieks, F., M. Piery and L. Quadriello. 1991. The natural nest of the Africanzied honey bee near Tapachula, Chiapas, Mexico. *Can. Entomol.* 123:353–359.

Rinderer, T. E. 1986. Africanized bees: The Africanization process and potential range in the United States. *Bull. Entomol. Soc. Amer.* 32:222–227.

Rinderer, T. E., S. M. Buco, W. L. Rubink, H. V. Daly, J. A. Stelzer, R. M. Riggio, and F. C. Bapista. 1993. Morphometric identification of Africanized and European honey bees using large reference populations. *Apidologie* 24(6):569–585.

Rinderer, T. E. and A. M. Collins. 1991. Foraging behavior and honey production. In: *The "African" Honey Bee*. Ed. M. Spivak, D. J. C. Fletcher and M. D. Breed. Westview Press, Boulder, CO. Ch. 12. 219–241.

Rinderer, T. E., B. P. Oldroyd, and W. S. Sheppard. 1993. Africanized bees in the U.S. *Scientific American* 269:84–90.

Rinderer, T. E., J. A. Stelzer, B. P. Oldroyd, S. M. Buco, & W. L. Rubink. 1991. Hybridization between European and Africanized honey bees in the neotropical Yucatan peninsula. *Science* 253:309–311.

Rinderer, T. E., H. A. Sylvester, M. A. Brown, J. D. Villa, D. Pesante, and A. M. Ccollins. 1986. Field and simplified techniques for identifying Africanized and European honey bees. *Apidologie* 17:33–48.

Rinderer, T. E., H. A. Sylvester, A. M. Ccollins, and D. G. Pesante. 1986. Identification of Africanized and European honey bees: effects of nurse-bee genotype and comb size. *Bull. Entomol. Soc. Amer.* 23(3):150–152.

Sheppard, W. S., T. E. Rinderer, J. A. Mazzoli, J. A. Stelzer and H. Shimanuki. 1991. Gene flow between African- and European-derived honey bee populations in Argentina. *Nature* 349:782-784.

Smith, D. R., O. R. Taylor and W. W. Brown. 1989. Neotropical Africanzied honey bees have African mitochondrial DNA. *Nature* 339:213–215.

Smith, D. R. 1991. Mitochondrial DNA and honey bee biogeography. In: *Diversity in the Genus Apis*. Ed. D. R. Smith. Westview Press, Boulder, CO. 131–176.

Spivak, M. 1991. The Africanization process in Costa Rica. In: *The "African" Honey Bee*. Ed. M. Spivak, D. J. C. Fletcher and M. D. Breed. Westview Press, Boulder, CO. Ch. 7 137–155.

Spivak, M. and E. Erickson. 1992. Do measurements of worker cell size reliably distinguish Africanized from European honey bees (*Apis mellifera* L.). *Amer. Bee Jour.* 132(4):252–255.

Spivak, M., D. J. C. Fletcher, & M. D. Breed. 1991. Introduction. In: *The "African" Honey Bee*. Ed. M. Spivak, D. J. C. Fletcher and M. D. Breed. Westview Press, Boulder, CO. 1–9.

Sylvester, H. A. and T. E. Rinderer 1987. Fast Africanized bee identification system (FABIS) manual. *Amer. Bee Jour.* 127:511–516.

Sylvester, H. A., T. E. Rinderer and H. Shimanuki. 1992. Certification options for dealing with Africanized bees. *Amer. Bee Jour.* 132(3):182–185.

Taylor, O. R. 1985. African bees: potential impact in the United States. *Bull. Entomol. Soc. Amer.* 31:15–24.

Taylor, O. R. 1999. Displacement of European honey bee (*Apis mellifera*) subspecies by an invading African subspecies in the Americas. *Proc. 35th Apimondia Internat. Congress*, Vancouver, Canada. 202.

Taylor, O. R. 1999. Displacement of European honey bee subspecies by an invading African subspecies in the Americas. *Proc. Apic. for 21st Century.* (Michigan State Univ.). Wicwas Press, Cheshire, CT. 38–46.

Taylor, O. R., A. Delgado, and F. Brizuela. 1991. Rapid loss of European traits from feral neotropical African honey bee population. *Amer. Bee Jour.* 131:783–784.

Taylor, O. R. and G. A. Rowell. 1988. Drone abundance, queen flight distance, and the neutral mating model for the honey bee, *Apis mellifera.* In: *Africanized Honey Bees and Bee Mites.* Ed. G. Needham, R. E. Page, Jr., M. Delfinado-Baker and C. Bowman. John Wiley & Sons, Inc., NY. 173–183.

Taylor, O. R. and G. B. Williamson. 1975. Current status of the Africanized bee in northern South America. *Amer. Bee Jour.* 115:92–94, 98.

Winston, M. L. 1992. *Killer Bees.* Harvard Univ. Press, Cambridge, MA. 162 pp.

Chapter 5

Anonymous. 1986. $8 million Africanized bee barrier proposed. *Amer. Bee Jour.* 126(11):729–734.

Boreham, M. M. and D. W. Roubik. 1987. Population change and control of Africanized honey bees (Hymenoptera: Apidae) in the Panama Canal area. *Bull. Entomol. Soc. Amer.* 33:34–39.

Burgett, M. D. and J. T. Ambrose. 1976. Biological control of the African-Brazilian honey bee. *Amer. Bee Jour.* 116(1):22, 26.

Dietz, Alfred. 1992. Honey bees of the world. In: *The Hive and the Honey Bee.* Dadant and Sons, Inc., Hamilton, IL. Ch. 2. 23–71.

Eischen, F. A. 1999. Varroa and Africanized bees: An update. *Amer. Bee Jour.* 139:594–595.

Espina, D. 1986. *Beekeeping of the Assassin Bees.* Inst. Tech. Costa Rica, Cartago, Costa Rica. 170 pp.

Flakus, Greg. 1993. *Living* with *Killer Bees.* Quick Trading Co., Oakland, CA. 144 pp.

Fottum, K. 1989. The Mexican connection. *Glean. Bee Cult.* 117(4):228, 230.

Hall, H. G. and K. Muralidharan 1989. Evidence from mitochondrial DNA that African honey bees spread as continuous maternal lineages. *Nature* 339:211–213.

Kempff Mercado, N. 1970. La apicultura en Bolivia. *Rev. Univ. Gabriel René Moreno* 27–28:6–15.

Kent, R. 1989. The African honey bee in Peru: An insect invader and its impact on beekeeping. *Appl. Geog.* 9:237–257. (reprinted with minor changes as "The Africanized bee in Peru." In: *The "African" Honey Bee*. Ed. M. Spivak, D. J. C. Fletcher and M. D. Breed. Westview Press, Boulder, CO. Ch. 19. 373–398.

Levin, M. D. 1974. Hybridization of honey bees in South America. *Bull. Entomol. Soc Amer.* 20(4):294–296.

Labougle, J. M. 1991. Beekeeping in Mexico: Past, present and future. *Amer. Bee Jour.* 131:132–135.

Loper, G. M., J. Fewell, D. R. Smith, W. S. Sheppard and H. Schiff. 1999. Genetic changes of a population of feral honey bees in the Sonoran desert of southern Arizona following the arrival of *Acarapis woodi, Varroa jacobsoni* and Africanization. In: *Proc. Apiculture for 21st Century*. (Michigan State Univ.) Wicwas Press, Cheshire, CT. 47–51.

National Academy of Sciences (NAS). 1972. *Final report. Committee on the African Honey Bee*. PB-211. 952. National Research Council, National Academy of Sciences. Tech Info. Service, U.S. Dept. of Commerce, Springfield, VA. 95 pp.

Nogueira-Neto P. 1964. The spread of the fierce African bee in Brazil. *Bee World* 45:119–121.

Page, R. E., Jr. 1998. Blessing or curse? Varroa mite impacts Africanized bee spread and beekeeping. *Calif. Agric.* 52(2):7–12.

Potter, A. 1977. *Killer Bees*. Grosset & Dunlap, NY. 159 pp.

Rinderer, T. E. 1986. Africanized bees: The Africanization process and potential range in the United States. *Bull. Entomol. Soc. Amer.* 32:222–227.

Rinderer, T. E., B. P. Oldroyd and W. S. Sheppard. 1993. Africanized bees in the U.S. *Scientific American* 269:52–58.

Rinderer, T. E., H. A. Sylvester, M. A. Brown, J. D. Villa, D. Pesante, and A. M. Collins. 1986. Field and simplified techniques for identifying Africanized and European honey bees. *Apidologie* 17:33–48.

Rinderer, T. E., J. E. Wright, H. Shimanuki, F. Parker, E. Erickson and W. T. Wilson. 1987. The proposed honey-bee regulated zone in Mexico. *Amer. Bee Jour.* 127:160–164.

Roubik, D. 1986. Long-term consequence of the Africanized honey bee invasion: implications for the United States. *Proc. Africanized Honey Bee Symp.* Amer. Farm Bur., Atlanta, GA. 46–52.

Rowell, M. S., L. A. Bradley and C. L. Cole 1993. Range extension in Texas. *Amer. Bee Jour.* 133(2):84–87.

Rubink, W. L., E. A. Sugden, P. Luevano-Martinez, A. M. Collins and W. T. Wilson. 1994. Subtropical honey-bee swarming dynamics before and after Africanization in northeastern Mexico and southern Texas. *Proc. 5th International Conf. Apic. Trop. Clim.* 186–192.

Rubink, W. L., W. T. Wilson, A. M. Collins, and P. Luevano-Martinez. 1995. Comparative Africanization rates in feral honey bee populations at three latitudes in northeastern Mexico and southern Texas. *Amer. Bee Jour.* 135(12):829.

Schiff, N. M., W. S. Sheppard, G. M. Loper and H. Shimanuki. 1994. Genetic diversity of feral honey bee (Hymenoptera: Apidae) populations in the southern United States. *Ann. Entomol. Soc. Amer.* 87(6):842–848.

Schmidt, J. O. and S. C. Thoenes 1987. Swarm traps for survey and control of Africanized honey bees. *Bull. Entomol.. Soc Amer.* 33:155–158.

Schmidt, J. O., S. C. Thoenes and R. Hurley. 1989. Swarm traps. *Amer. Bee Jour.* 129(7):468–471.

Schmidt, J. O. and S. C. Thoenes 1990. Honey bee (Hymenoptera: Apidae) preferences among artificial nest cavities. *Ann. Entomol. Soc. Amer.* 83(2):271–274.

Schmidt, J. O. 1995. Biological control of Africanized honey bees with swarm traps: economics and logistics. *Amer. Bee Jour.* 135(12):829–830.

Schneider, S. and R. Blyther. 1988. The habitat and nesting biology of the African honeybee *Apis mellifera scutellata* in the Okavango River delta, Botswana, Africa. *Insectes Soc.* 35:167–181.

Sheppard, W. S. 1989. A history of the introduction of honey bee races into the United States. *Amer. Bee Jour.* 129:664–667.

Sheppard, W. S., T. E. Rinderer, J. A. Mazzoli, J. A. Stelzer and H. Shimanuki. 1991. Gene flow between African- and European-derived honey bee populations in Argentina. *Nature* 349:782–784.

Sheppard, W. S., A. E. E. Soares, D. De Jong, and H. Shimanuki. 1991. Hybrid status of honey bee populations near the historic origin of Africanization in Brazil. *Apidologie* 22:643–652.

Smith. F. G. 1960. *Beekeeping in the Tropics.* Longmans, London.

Smith, F. G. 1961. The races of honeybees in Africa. *Bee World* 42:255–260.

Smith, D. R., O. R. Taylor, and W. W. Brown. 1989. Neotropical Africanized honey bees have African mitochondrial DNA. *Nature* 339:213–215.

Spivak, M., D. J. C. Fletcher, & M. D. Breed. 1991. Introduction. In: *The "African" Honey Bee.* Ed. M. Spivak, D. J. C. Fletcher and M. D. Breed. Westview Press, Boulder, CO. Introduction. 1–9.

Sugden, E. A. and K. R. Williams. 1991. October 15: the day the bee arrived. *Glean. Bee Cult.* 119(1):18–21.

Sugden, E. A., A. M. Collins, W. L. Rubink, L. A. Bradley, K. Davis and E. Quintero. 1994. Two years of "Africanization"; honey bee stinging incidents in Texas. *BeeScience* 3(2):68–75.

Taber, S. 1977. The African bee in Louisiana. *Amer. Bee Jour.* 117(3):152–154.

Taylor, O. R. 1977. The past and possible future spread of Africanized honeybees in the Americas. *Bee World* 58:19–30.

Taylor, O. R. 1985. African bees: potential impact in the United States. *Bull. Entomol. Soc. Amer.* 31:15–24.

Taylor, O. R. 1988. Ecology and economic impact of African and Africanized honey bees. In: *Africanized Honey Bees and Bee Mites.* Ed. G. R. Needham, R. E. Page, M. Delfinado-Baker, and C. Bowman. Ellis Horwood Ltd, Chichester, England. Ch. 3. 29–40.

Tew, J. E., C. H. Bare and J. D. Villa 1988 The honey bee regulated zone in Mexico. *Amer. Bee Jour.* 128:673–675.

Texas Extension. 1991. Africanized honey bee alert. *Texas Extension Leaflet Series.*

Vaissiere, B. E. and F. G. Maxwell 1989. Potential impct of the Africanized honey bees in Texas. *Texas Afr. Honey Bee Program.* Mimeo 27 pp.

Villa, J. D., N. Koeniger and T. E. Rinderer. 1991. Overwintering of Africanized, European and hybrid honey bees in Germany. *Environ. Entomol.* 20(1):39–43.

Wiese, H. 1977. Apiculture with Africanized bees in Brazil. In: *African Bees.* Ed. D. J. C. Fletcher. Apidmondia, Pretoria. 67–76. (Also in *Amer. Bee Jour.* 117:166–170).

Wiese, H. 1985. *Nova apicultura 6th ed.* São Paulo, Brazil. 362 pp.

Winston, M. L. 1979. The potential impact of the Africanized honey bee on apiculture in Mexico and Central America. *Amer. Bee Jour.* 119:584–585 and 642–644.

Winston, M. L. 1988. The impact of a tropical-evolved honey bee in temperate climates of North America. In: *Africanized Honey Bees and Bee Mites.* Ed. G. R. Needham, R. E. Page, M. Delfinado-Baker, and C. Bowman. Ellis Horwood Ltd, Chichester, England. Ch. 16. 135–140.

Winston, M. 1992. *Killer Bees.* Harvard Univ. Press, Cambridge, MA. 162 pp.

Woyke, J. 1973. Experiences with *Apis mellifera adansonii* in Brazil and Poland. *Apiacta* 8:115–116.

Chapter 6

Boreham, M. M. and D. W. Roubik. 1987. Population change and control of Africanized honey bees (Hymenoptera: Apidae) in the Panama Canal area. *Bull. Entomol. Soc. Amer.* 33:34–39.

Boreham, M. M. and D. W. Roubik 1990. Learning to live with Africanized honey bees. *Interciencia* 15(3):146–153.

Buchmann, S. L. 1982. Africanized bees confirmed in Panama. *Amer. Bee Jour.* 122:322.

Caron, D. M. 1984. The McClymonts of Panama. *Glean. Bee Cult.* 112(11):607–608.

Caron, D. M. 1987. Surviving with Africanized bees. *Amer. Bee Jour.* 127(8):561–563.

Caron, D. M. 1988. Panama. *Glean. Bee Cult.* 117(12):675, 677–678.

Caron, D. M. and J. Camargo. 1995. Rebuilding beekeeping in Panama. *Amer. Bee Jour.* 135:243–245.

Caron, D. M. and B. Gray. 1991. The impact of the Africanized bee on beekeeping in Panama. *BeeScience* 1(3):139–143.

Caron, D. M. and John Reese 1982. An entomological visit to Panama. *Amer. Bee Jour.* 122(2):112–114.

Cordovez, E. 1986. The handling of the Africanized honeybee by the Panama Canal Commission. *Proc. African Honey Bee Symposium*. American Farm Bureau, Atlanta, GA. 36–45.

Crane, E. 1976. The exploitation of bees in tropical environments. In: *Apiculture in Tropical Climates*. Ed. E. Crane. IBRA, Cardiff, Wales. 5–7.

De Jong, D. 1994. The African experience. *Bee Culture* (8):453, 456–461.

Dressler, W. 1976. The use of movable-frame hives in development programmes (Africa and Latin America). In: *Apiculture in Tropical Climates*. Ed. E. Crane. IBRA, Cardiff, Wales. 119–124.

Espina, D. 1986. *Beekeeping of the Assassin Bees*. Inst. Tech. Costa Rica, Cartago, Costa Rica. 170 pp.

Kempe, F. 1990. *Divorcing the dictator—Americas bungled affair with Noreiga*. Putnam & Sons, NY. 469 pp.

Fore, T. 1986. Diverse group visits Panama to see African bees. *Speedy Bee* Mar '86:13–16 and Apr '86:15–18.

McClymont, S. 1969. Beekeeping in Panama. *Glean. Bee Cult.* 97(1):24–27.

Petersen, H, 1976. German beekeeping development programme in Central America. In: *Apiculture in Tropical Climates*. Ed. E. Crane. IBRA, Cardiff, Wales. 155–158.

Roubik, D. W. 1982. Report and preliminary recommendations concerning the Africanized honey bee in Panama. *Speedy Bee* June '82. 12–13.

Urena, A. P. 1982. La abeja Africada: un insecto muy calumniado. *Sucros* 1(9):4–5.

Winston, M. L. 1992. *Killer Bees*. Harvard Univ. Press, Cambridge, MA. 162 pp.

Chapter 7

Boreham, M. M. and D. W. Roubik. 1987. Population change and control of Africanized honey bees (Hymenoptera: Apidae) in the Panama Canal area. *Bull. Entomol. Soc. Amer.* 33:34–39.

Boreham, M. M. and D. W. Roubik 1990. Learning to live with Africanized Honey Bees. *Interciencia* 15(3):146–153.

Calderon, R. A., A. Ortiz, B. A. Parisio and M. T. Ruiz. 2000. Varroa in Panama: detection, spread and prospects. *Bee World* 81(3):126–128.

Caron, D. M. 1984. The McClymonts of Panama. *Glean. Bee Cult.* 112(11):607–608.

Caron, D. M. 1985. Entomological society holds Africanized bee symposia. *Amer. Bee Jour.* 125(3):185–187.

Caron, D. M. 1987. Surviving with Africanized Bees. *Amer. Bee Jour.* 127(8):561–563.

Caron, D. M. 1988. Panama. *Glean. Bee Cult.* 116(6):333–334.

Caron, D. M. 1989. Panama's perils. *Glean. Bee Cult.* 117(12):675, 677–678.

Caron, D. M. 1994. Take a trip. *Glean. Bee Cult.* 122(10):561.

Caron D. M. and B. Gray. 1991. The impact of the Africanized bee on beekeeping in Panama. *BeeScience* 1(3):139–143.

Caron, D. M. and J. Camargo. 1995. Rebuilding beekeeping in Panama. *Amer. Bee Jour.* 135:243–245.

Cobey, S. and S. Locke. 1986. The Africanized bee: a tour of Central America. *Amer. Bee Jour.* 126(6):434–440.

Collins, A. M., W. L. Rubink, J. I. Cuadriello-Aguilar and R. L. Hellmich. 1994. Use of insect repellents to disperse defending honey bees. *Proc. 5th Conf. Apic. Trop. Climates: Trinidad & Tobago.* IBRA,Cardiff, Wales. 17–20.

Cordovez E. 1986. The handling of the Africanized honeybee by the Panama Canal Commission. *Proc. Africian Honey Bee Symposium.* American Farm Bureau, Atlanta, GA. 36–45.

Fore, T. 1986. Diverse group visits Panama to see African bees. *Speedy Bee* Mar '86:13–16 and Apr '86:15–18.

Hellmich, R. L. and T. E. Rinderer. 1992. Beekeeping in Venezuela. In: *The "African" Honey Bee.* Ed. M. Spivak, D. J. C. Fletcher and M. D. Breed. Westview Press, Boulder, CO. Ch. 20. 399–411

Kent, R. B. 1978. *Apicultural development in Central America and Panama: Some historic and economic considerations.* Mimeo (source unknown): 9 pp.

Levin, M. D. 1978. *Recent observations on Africanized honey bees in South and Central America.* Trip report. USDA mimeo. 7 pp.

MIDA. 1984. *Technicas de Manejo de Abeja Africanizada.* Panama MIDA mimeo. 5 pp.

MIDA. 1985 *La Abeja Africanizada en Panama*. Panama MIDA mimeo. 11 pp.

Moore, W. S. 1982. Biases of information about bees and beekeepers in popular magazines. *Amer. Bee Jour.* 105:405–406.

Mondolfi, A. 1983. *Recomendaciones Basicas en el Tratamiento de Picadas de Abejas*. Panama MIDA mimeo. 4 pp.

Nunnamaker, R. A. 1979. Being stung by the press. *Amer. Bee Jour.* 119:587–591, 646–647, 657.

Otis, G. W. 1982. Population biology of the Africanized honey bee. In: *Social Insects of the Tropics, Vol 1*. Ed P. Jaisson. Univ. Paris-Nord, Paris. 209–219.

Pardo, A. M. 1984. *La abeja Africanizada—algunos aspectos sobre su origen biologia y manejo*. Panama MIDA leaflet. 61 pp.

Roubik, D. W. 1982. Report and preliminary recommendations concern ing the Africanized honey bee in Panama. *Speedy Bee* June '82:12–13.

Roubik, D., R. J. Schmalzel and J. E. Moreno. 1984. Estudio apibotanico de Panama: Cosecha y fuentes de polen y nectar usados por *Apis mellifera* y sus patrones estaciones y anuales. *Organismo Internat. Regional de Sanidad Agropecuaria. (ORISA) Tech. Bull. 24*. San Salvador, El Salvador. 73 pp.

Stoner, A. and W. T. Wilson. 1972. A review of the public's response to Africanized honey bees. *Glean. Bee Cult.* 100(9):405–408.

Tomasko, Maryann and D. M. Caron. 1994. Bee successful in Panama. *Amer. Bee Jour.* 134(1):47–49.

Vanegas, R.. 1985. *Seminario de capacitacion para particpantes en captura y eliminacion de enjambres de abejas Africanizada*. Panama MIDA Mimeo. 6 pp.

Willenjohn, C. 1996. Escape from killer bees. *Soldier* 51(7):20.

Chapter 8

Arce Arce, H. G., J. W. Van Veen, and M. J. Sommeneijer. 1992. Beekeeping with stingless bees (Hymenoptera: Apidae: Meliponinae) in Costa Rica. *Proc. 5th Inter. Conf. Apic. Trop. Clim.* IBRA, Cardiff, Wales. 209–213.

Arce Arce, H. G. and J. W. Van Veen. 1997. Perspectives for honey production in the tropics. *Proc. Nectar Symp.*, Univ. of Utrecht, Netherlands. 103–115.

Calkins, C. C. 1969. Notes on beekeeping in Costa Rica. *Glean. Bee Cult.* 97:526–530.

Calvo-Pineda, E. 1988. *Censo Agropecuaria*. Dir. Censos, Costa Rica. 20 pp.

Cark, B 1989. We can work with Africanized bees. *Speedy Bee* 7–8.

Caron, D. M. 1994. Take a trip. *Bee Culture* 122(10):560.

Caron, D. M. 2001 (anticipated). Neotropical Pollinators: Crises, Crossroads and Conservation. Thomas Say. *Ent. Soc. Amer.* Pub. Ch. 5.

Caron, D. M. and B. Gray. 1991. The impact of the Africanized bee on beekeeping in Panama. *BeeScience* 1(3):139–143.

Caron, D. M. and J. Camargo. 1995. Rebuilding beekeeping in Panama. *Amer. Bee Jour.* 135:243–245.

Cobey, S. and S. Locke. 1986. The Africanized bee: A tour of Central America. *Amer. Bee Jour.* 126(6):434–440.

Crane, Eva 1992. Stingless bees in Costa Rica. *Beekeeping and Development.* IBRA, Cardiff, Wales. 21:6.

Danka, R. G., G. M. Loper, J. D. Villa, J. L. Williams, E. A. Sugden, A. M. Collins and T. E. Rinderer. 1994. Abating feral Africanized honey bees to enhance mating control of European queens. *Apidologie* 25:520–529.

Danka, G., J. D. Villa and N. E. Gary. 1992. Comparative foraging distances of Africanized, European and hybrid honey bees during cantalope pollination. *Amer. Bee Jour.* 132(12):806.

Dyce, E. J. 1953. Beekeeping in Costa Rica. *Amer. Bee Jour.* 96(7):296–298.

Dyer, D. 1989. Killing the killer bees. *Tico Times* Sept 8:10–11.

Erickson, E. H., B. J. Erickson and A. M. Young. 1986. Management strategies for Africanized honey bees: concepts strengthened by our experiences in Costa Rica. *Glean. Bee Cult.* 114:456–457, 459, 506–507, 534.

Espina, D. 1986. *Beekeeping of the Assassin Bees.* Inst. Tech. Costa Rica, Cartago, Costa Rica. 170 pp.

Espina, D and G. S. Ordetx 1984. *Apicultural Tropical.* Inst. Tech. Costa Rica. Cartago, Costa Rica. 506 pp

Fore, T. 1986. Diverse group visits Panama to see African bees. *Speedy Bee* March:13–16, April:15–18.

Hall, H. G. and K. Muralidharan 1989. Evidence from mitochondrial DNA that African honey bees spread as continuous maternal lineages. *Nature* 339:211–213.

Flakus, G. 1993. *Living* with *Killer Bees.* Quick Trading Co., Oakland, CA. 144 pp. (Available from A. I. Root Co.).

Gonzales, M. 1989. *Technicas modernas para la explotacion de la abeja Africanizada.* APES, Alajuela, Costa Rica. 70 pp.

Kent, R. B. 1978. Apicultural development in Central America and Panama: some historic and economic considerations. Mimeo (source unknown). 9 pp.

Kent, R. B. 1979. Diversidad ecologica y las regiones apicolas de Costa Rica. *Rev. Geografica* 90:65–95.

Kent, R. B. 1979. Technical and financial aspects of fixed comb and movable frame beekeeping in Costa Rica. *Amer. Bee Jour.* 122:36–38, 43, 127–128, 135.

Kent, R. B. 1984. Honeybees and the economic development of beekeeping in Costa Rica. *Ibero-Amerikanisches Archiv.* 10:323–329.

Levin, M. D. 1978. *Recent observations on Africanized honey bees in South and Central America.* Trip report. USDA mimeo 7 pp.

Lobo, J. A. 1995. Morphometric, isozymic and mitochondrial variability of Africanized honeybees in Costa Rica. *Heredity* 75:133–141.

Loper, G. M. and E. A. Sugden 1992. Local Africanized honey bee abatement using drones as insecticide vectors in Costa Rica. *Amer. Bee Jour.* 132(12):810–811.

Petersen, H, 1976. German beekeeping development program in Central America. In: *Apiculture in Tropical Climates.* Ed. E. Crane. IBRA, Cardiff, Wales. 155–158.

Ramirez, W. 1992. Ten years of Africanization in Costa Rica, the experience of a beekeeper. *Amer. Bee Jour.* 132 (12):785.

Ramirez, W. 1993. What you should know about African bees now that they have arrived in the U.S.A. *Amer. Bee Jour.* 133(1):51–52.

Ramirez, W. 1993. Swarming of the African bees in the central valley of Costa Rica. *Amer. Bee Jour.* 133:283–284.

Ramirez, W. 1993 *Truths and myths about the African bee (AHB) in the new world.* Author's mimeo 4 pp.

Ramirez, W. 2001. Management of commercial colonies of Africanized honey bees in Costa Rica. *Proc. 2nd Inter. Conf. on Africanized Bees and Bee Mites.* To be published by A. I. Root, Co., Medina, OH.

Roubik, D. W. 1995. *Pollination of Cultivated Plants in the Tropics.* FAO Agric. Bull. *118.* 198 pp.

Sommeijer, M. J. 1996. A regional program for training and research on tropical beekeeping and tropical bees in Costa Rica. *Bee World* 77(1):3–7.

Spivak, M, S. Batra, F. Segredea, A. L. Castro and W. Ramirez. 1989. Honey production of Africanized and European honey bees in Costa Rica. *Apidologie* 20:207–220.

Spivak, M. 1991. The Africanization process in Costa Rica. In: *The "African" Honey Bee.* Ed. M. Spivak, D .J. C. Fletcher and M. D. Breed. Westview Press, Boulder, CO. Ch. 7. 137–155.

Spivak, M. and E. Erickson. 1992. Do measurements of worker cell size reliably distinguish Africanized from European honey bees (*Apis mellifera* L.). *Amer. Bee Jour.* 132(4):252–255.

Taylor, O. R. 1984. Challenges Africanized bee article. *Amer. Bee Jour.* 124(5):395–396.

Taylor, O. R. 1984. African bees in Costa Rica; predicted to reach Texas by 1989. *Amer. Bee Jour.* 124(6):475–476.

Tipton, J. 1995. Those bees and beekeepers in Brazil. *Amer. Bee Jour.* 135:331–335.

Tomasko, M. and D. M. Caron 1991. The African bee—not all bad news for Costa Rica. *Amer. Bee Jour.* 131(1):25–28.

Tomasko, M. and D. M. Caron. 1994. Bee successful in Panama. *Amer. Bee Jour.* 134(1):47–49.

Villa, J. D., T. E. Rinderer and A. M. Collins. 1993. "Overwintering" of Africanized, European and hybrid honey bees (Hymenoptera: Apidae) in the Andes of Venezuela. *Environ. Entomol.* 22(1):183–189.

Winston, M. L. 1979. The potential impact of the Africanized honey bee on apiculture in Mexico and Central America. *Amer. Bee Jour.* 119:584–585 and 642–644.

Chapter 9

Anonymous. 1992. *Coca production in the Chapare region of Bolivia.* Mimeo 5 pp.

Apisbol. 1995. La miel de abejas, cristalizada es pura. *Informativo Afiche.* Apisbol, Cochabamba, Bolivia. Mimeo 4 pp.

Ardaya, R. I., J. E. Villaroel and C. B. Mariscal. 1995. Situacion del sector forestal del Department de Cochabamba. *Plan Acción Forestal Bolivia.* La Paz, Bolivia. 106 pp.

Arispe, O. 1988. *Manual Practico de Iniciacion Apicola.* Apisbol, Cochabamba, Bolivia. 60 pp.

Blanes, J. 1983. Agricultura campesina y los mercardos de alimentos. *Ceres,* Cochabamba, Bolivia. 121 pp.

Caron, D. M. 1995. *Guia practica—Apicultura.* CEDEAGRO, Cochabamba, Bolivia. Mimeo. 40 pp

Caron, D. M. 1995. *Pollination requirements of alternative crops grown in the Chapare.* CORDEP-DAI mimeo. 15 pp.

Caron, D. M. 1996. Bolivia honey festival. *Amer. Bee Jour.* 136:515–516.

Caron, D. M. 1996. Mizque Valley Bolivia—example of an effective rural beekeeping project. *Amer. Bee Jour.* 136(6):407–409.

Caron, D.. M. 1996. Peace Corps volunteer in Bolivia. *Bee Culture* 124(10):567–569.

Caron, D. M. 1996. Is beekeeping an alternative to coca growing? *BeeScience* 4(3):84–86.

Caron, D. M. 1997. Crises to crossroads—Africanized bees in the Americas. *Part. Amer. Directions* June/July 1997. 37–40.

Cosenza, G. W. 1972. Estudio dos exames de migraçae de abelhas Africans. *Proc 1º Braz. Congr. Apiculture.* Santa Catarina, Brazil. 128–129.

Dietz, A. and C. Vargara 1995. Africanized honey bees in temperate climates. *Bee World* 76:56–71.

Duay, P. and D. De Jong. 1994. Variacion de la produccion de miel con abejas Africanizadas seleccionadas (*Apis mellifera scutellata*), no seleccionada and híbridas (*A. m. lingustica* x *A. m. scutellata*) en condiciones de campo en Brasil. *Proc. IV Cong. Apiculture Iberolatinoamericano*. Rio Cuarto, Cordoba, Argentina. Com I. 7–10.

FAO. 1995. Proyecto apoyo al manejo, conservacion y explotacion de los recursos forestales en el trópico de Cochabamba, Bolivia. FAO Leaflet, Chimore, Bolivia. 2 pp.

Guaraguara, V. 1995. *Perfel de proyecto para fortalecer el desarrollo apicola en el valle de Cochabamba*. Fundforma, Cochabamba, Bolivia.. Mimeo 7 pp.

Hofman, C. 1993. *Impacto del crédito agricola en la unidad productos familiar del Chapare*. CORDEP-DAI, Cochabamba, Bolivia. 12 pp

INE. 1993. *Programa de desarollo alternativo regional. La migracion de la poblacion en el Chapare. Documento elaborado en base a resultada censo de poblacion y vivienda 1992*. INE, La Paz, Bolivia 50 pp.

INE/PDAR/CORDEP-DAI 1994. *Agricultural survey of the tropical area of the Department of Cochabamba*. INE/PDAR/CORDEP-DAI, Cochabamba, Bolivia. 106 pp.

Kempff Mercado, N. 1957. Control de las abejas indigenas indeseables. *Gaceta del Colmenar* 19:89–90.

Kempff Mercado, N. 1966. Abejas indigenas—su explotacion racional. *Rev. Univ. Gabiela René Moreno* 23–24:162–165.

Kempff Mercado, N. 1970. La apicultura en Bolivia. *Rev. Univ. Gabriel René Moreno* 27–28:6–15.

Kempff Mercado, N. 1972. Abejas Africans—contribucion a su conociemiento. *Rev. Univ. Gabriel René Moreno* 34:3–15.

Kempff Mercado, N. 1973. The African bees: contibution to their knowledge. *Apiacta* 8:121–126.

Kempff Mercado, N. 1980. *Flora apicola subtropical de Bolivia*. Univ. Gabriel René Moreno. 77 pp.

Kent, R. 1990. *Regional planning and strategy analysis for the Chapare Regional Development Project*. DAI, Washington, DC. 29 pp.

Laserna, R. 1994. *Las drogas y el ajuste en Bolivia*. CEDLA, Cochabamba, Bolivia. 18 pp.

Mangum, W. A. 2000. Adventures in Bolivian beekeeping. *Amer. Bee Jour.* 140(10):767–770.

Michener, C. D. 1975. The Brazilian bee problem. *Ann. Rev. Entomol.* 20:239–416.

Munro, J. A. 1953. Beekeeping in Bolivia. *Glean. Bee Cult.* 81:204–208.

Pillco, P. 1990. *Apicultura 1 & 2*. Proyecto Agroyungas, Coroico, Bolivia. Mimeo 49 and 48 pp.

Pinto, F. A. and R. A. Trujillo. 1995. *Flores de interes apicola y poliniza ción de cultivos.* Thesis UMSS Escuela Technica Superior de Agronomia. Cochabamba, Bolivia. 87 pp.

Shedd, C. M. 1949. Bees in Bolivia. *Glean. Bee Cult.* 77:273–274.

Stearman, A. M. 1981. Working the "Africans" in Eastern Bolivia. *Amer. Bee Jour.* 121:28–35, 43–44.

USAID. 1992. *Program objectives and action plan (1993-1997).* DAI, Bethesda, MD 90 pp.

USAID. 1992. *Coca production in the Chapare Region of Bolivia.* Mimeo 5 pp.

USAID. 1995. *Chapare—progress of economic development.* USAID leaflet. Cochabamba, Bolivia. 9 pp.

USAID. 1995. *El Nuevo Chapare.* PDAR Newsletter 7 (Oct 1995):5.

Chapter 10

Belize

Burgett, D. M., G. Fisher, and R. O'Doherty. 1990. The Africanization of a nation: Belize, Central America. *Glean. Bee. Cult.* 118:534–537.

Burgett, D. M., & G. Fisher. 1990. Final report: *A review of the Belizean honey bee industry.* Ore. State Univ. Mimeo 30 pp.

Caron, D. M. 1991. Meet Tom Hyden—Peace Corps beekeeping volunteer in Belize. *Amer. Bee Jour.* 131:561–563.

Hyden, T. 1994. Belize. *IAAD Cornucopia Newsletter* 5:2.

Hyden, T. 1991. The Southern Beekeepers Co-op Society, Belize, C.A. *IAAD Cornucopia Newsletter* 5:7–8.

Krell, R. 1992. The results of diminishing forest resources on management and productivity of honey bees based on examples of three tropical countries. *Proc. 5th Inter. Conf. Apiculture Trop. Climates.* IBRA, Cardiff, Wales. 35–37.

Mayer, D. F., and M. Cal. 1988. New beekeeping industry in Belize faces many of the same problems we have. *Amer. Bee Jour.* 128:197–198.

Mulzac, H. C. 1979. Beekeeping in Belize. In: *Beekeeping in Rural Development.* IBRA, Cardiff, Wales. 157–164.

Poovey, C. 1992. Royal Hayan bee. *Beekeeping and Development.* IBRA, Cardiff, Wales. 25:6–7.

Sylvester, H. A., and T. E. Rinderer. 1987. Fast Africanized bee indentification system (FABIS) manual. *Amer. Bee Jour.* 127:511–516.

Villanueva, R. 1994. Nectar sources of European and Africanized honey bee (*Apis mellifera* L) in the Yucatan Peninsula, Mexico. *Jour. Apic. Res.* 33:44–58.

Villanueva, R., A. Cesar-Dachary, and A. Sanchez-Vazquez. 1991. The impact of the Africanized honey bee in north Belize. *Amer. Bee Jour.* 131:183–186.

Brazil

Amarel, E. 1957 Honey bee activities and honey plants in Brazil. *Amer. Bee Jour.* 97:394–395.

Camazine, S., and R. A. Morse. 1988. The Africanized honeybee. *American Scientist* 76:465–471.

Cosenza, G. W. 1972. Estudio dos exames de migraçae de abelhas africans. *Proc. 1º Braz. Congr. Apiculture*, Santa Catarina, Brazil. 128–129.

De Jong, D. 1994. The African experience. *Bee Culture* 122:453, 456–461.

De Jong, D. 1996. AHB in Brazil, forty years of adaptation and success. *Bee World* 77(2):67–70.

Diniz, N. M., A. E. E. Soares and V. B. Pecci. 1994. Africanized honey bee control program in Ribeirão Preto City, São Paulo, Brazil. *Amer. Bee Jour.* 134:746–748.

Duay, P. and D. De Jong. 1994. Variacion de la produccion de miel con abejas Africanizadas seleccionadas (*Apis mellifera scutellata*), no seleccionada and hibridas (*A. m. lingustica* x *A. m. scutellata*) en condiciones de campo en Brasil. *Proc. IV Cong. Apiculture Iberolatinoamericano*. Rio Cuarto, Cordoba, Argentina. Com I. 7–10.

Eischen, F. 1994. "Beeless" beekeeping and other smart choices. *Amer. Bee Jour.* 134:664–665.

Gonçalves, L. S. 1973–4. The introduction of the African bees (*Apis mel lifera adansonii*) into Brazil and some comments on their spread in South America. *Amer. Bee Jour.* 114:414–415, 419, 448–450 and 115:8–10, 24.

Gonçalves, L. S. 1999. Biological impact caused by the Africanization of honey bees in Brazil and the future of beekeeping. *Proc. 36th Congress Apimondia*. Vancouver, Canada. 204

Gonçalves, L. S., A. C. Stort, and D. De Jong. 1991. Beekeeping in Brazil. In: *The "African" Honey Bee.* Ed. M. Spivak, D. J. C. Fletcher and M. D. Breed. Westview Press, Boulder, CO. Ch. 18. 329–372.

Gonçalves, L. S., W. E. Kerr, J. C. Netto and A. C. Stort. 1974. Some comments on the *"Final Report of the Committee on the African Honeybee."* Cornell Univ. Mimeo. 35 pp.

Lourenço, T. A. C. 1982. In spite of the African bees beekeeping prospers in Brazil. *Amer. Bee Jour.* 122:799–800.

Michener, C. D. 1975. The Brazilian bee problem. *Ann. Rev. Entomol.* 20:339–416.

National Academy of Sciences (NAS). 1972. *Final Report. Committee on the African Honey Bee.* PB-211. 952 National Research Council, National Academy of Sciences. Tech. Info. Service, U. S. Dept. of Commerce, Springfield, VA. 95 pp.

Nogueira-Neto, P. 1964. The spread of the fierce African bee in Brazil. *Bee World* 45:119–121.

Nogueira-Neto, P. 1972. Notas sobre a historia da apicultura Brasiliera. In: *Manual de Apicultura*. Ed. J. M. F. de Camargo. *Edit. Agron. Ceres*, São Paulo, Brazil. 17–32.

Nunamaker, R. P. 1979. Being stung by the press. *Amer. Bee Jour.* 119:587–591, 646–647 and 657.

Patty, G. E. 1978. *The honey industry of Mexico situation and prospects.* USDA For. Agri. Ser. M–285. 18 pp.

Rinderer, T. E., and A. M. Collins. 1991. Foraging behavior and honey production. In: *The "African" Honey Bee*. Ed. M. Spivak, D. J C. Fletcher and M. D. Breed. Westview Press, Boulder, CO. Ch. 12. 219–241.

Robinson, G. 1980. The potential for apicultural development in the third world. *Amer. Bee Jour.* 120:398–400.

Sanford, M. T. 1996. What of the African honey bee. *APIS Newsletter* 14:1–4.

Stoner, A., and W. T. Wilson. 1977. A review of the public's response to Africanized honey bees. *Glean. Bee Cult.* 405–408.

Sousa, R. M., B. M. Freitos, Z. B. Arafijo, and A. E. E. Soares. 2001. Seasonal changes in the Africanized honey bee (*Apis mellifera* L) population of the Coatinga vegetation in NE Brazil, *Proc. 2nd Int. Conf. Afr. Bees & Bee Mites*, Tucson, AZ.

Wiese, H. 1977. Apiculture with Africanized bees in Brazil. In: *African Bees* Ed. D. J. C. Fletcher. Apidmondia, Pretoria. 67–76. (also in *Amer. Bee Jour.* 117:166–170).

Winston, M. L. 1992. *Killer Bees.* Harvard Univ. Press, Cambridge, MA 162 pp.

Woyke, J. 1969. African honey bees in Brazil. *Amer. Bee Jour.* 109:342–344.

Peru

Harper, J. D. 1982. Africanized bees spread to Ecuador. *Speedy Bee* 11:7–8.

Kempff Mercado, N. 1972. Abejas Africanas—contribucion a su conociemiento. *Rev. Univ. Gabriel René Moreno* 34:3–15.

Kent, R. 1984. The African honey bee in Peru: An insect invader and its impact on beekeeping. *Appl. Geog.* 9:237–257.

Kent, R. B. 1989. The Africanized bee in Peru. In: *The "African" Honey Bee*. Ed. M. Spivak, D. J. C. Fletcher and M. D. Breed. Westview Press, Boulder, CO. Ch. 19. 373–398.

Spivak, M. 1991. The Africanization process in Costa Rica. In: *The "African" Honey Bee*. Ed. M. Spivak, D. J. C. Fletcher and M. D. Breed. Westview Press, Boulder, CO. Ch. 7. 137–156.

Trinidad and Tobago

Caron, D. M. 1989. Beekeeping in Trinidad. *Amer. Bee Jour.* 129:589–590.

Hallin, M. K. F. 1985. Farming aggressive honey bees. *T & T Agr. Ext. Bull. 5.* 6 pp.

Hallin, M. K. F. 1992. Apiculture in Trinidad with Africanized honey bees. *Proc. 5th Inter. Conf. Apic. Trop. Climates.* IBRA, Cardiff, Wales. 267–270.

Venezuela

Benson, K. 1985. American and Canadian beekeepers study Africanized bees in Venezuela. *Amer. Bee Jour.* 125:188–191.

Caron, D. M. 1985. Entomological society holds Africanized bee symposium. *Amer. Bee Jour.* 125:185–187.

Hellmich, R. L. and T. E. Rinderer. 1991. Beekeeping in Venezuela. In: *The "African" Honey Bee.* Ed. M. Spivak, D. J. C. Fletcher and M. D. Breed. Westview Press, Boulder, CO. Ch. 20. 399–411.

Levin, M. D. 1978. *Recent observation on Africanized honey bees in South and Central America.* Trip report. USDA mimeo. 7 pp.

Rinderer, T. E. 1985. Africanized honey bees in Venezuela: Honey pro--duction and foraging behavior. In: *Apiculture in Tropical Climates.* Vol. 3. IBRA, Cardiff, Wales. 112–116.

Rodriquez, R. G. 1978. Principios de manejo de apiarios durante el pre flujo en condiciones de Africanization. *Rep. Venezuela Min. Agric. y Cria* Bull. 3. 14 pp.

Rodriquez, R. G. 1986. *Manejo de la Abeja Africanizada.* Dir. Gen. Desarrollo Ganadero. Caracas, Venezuela. 280 pp.

Taylor, O. R. 1978. Observations on Africanized honey bees reported to South and Central American governing agencies. *Bull. Entomol. Soc. Amer.* 24:412-414

Chapter 11

Anderson, R. H., B. Buys, M. F. Jojannsmeier. 1983. *Beekeeping in South Africa.* Pretoria Bull. Dept. of Agric., No 394. 207 pp.

Andere, C., M. A. Palacio, E. Rodriguez, M. T. Dominquez, E. Figini and E. Bedascarrasbure. 2000. Evaluation of honey bee defensive behavior in Argentina—a field method. *Amer. Bee Jour.* 140:975–978.

Boreham, M. M. and D. W. Roubik 1990. Learning to live with Africanized honeybees. *Interciencia* 15(3):146–153.

Brandeburgo, M. A. 1990. Aggressive behavior of bees. *Ciencia y Cultura* 42(12):1025–1034.

Breed. M. D. 1991. Defensive behavior. In: *The "African" Honey Bee.* Ed. M. Spivak, D. J. C. Fletcher and M. D. Breed. Westview Press, Boulder, CO. Ch. 15. 299–308.

Breed. M. D., G. E. Robinson and R. E. Page. 1990. Division of Labor during honey bee colony defense. *Behav. Ecol. Sociobiol.* 27:395–401.

Boch, R. and W. C. Rothenbuhler. 1974. Defensive behavior and production of alarm pheromone in honey bees. *Jour. Apic. Res.* 13:217–221.

Brandeburgo, M. A. 1990. Aggressive behavior of bees. *Ciencia e Cultura* 42(12):1025–1034.

Collins, A. M., T. E. Rinderer, K. W. Tucker, H. A. Sylvester and J. J. Lackett. 1980. A model of honey bee defensive behavior. *Jour. Apic. Res.* 19:224–231.

Collins, A. M., T. E. Rinderer, J. R. Harbo, and A. B. Bolten. 1982. Colony defense by Africanized and European honey bees. *Science* 218:72–74.

Collins, A. M. 1985. Africanized honeybees in Venezuela: defensive behavior. *Proc. 3rd Int. Conf. Apic. Trop. Climates.* IBRA, Cardiff, Wales. 117–122.

Collins, A. M. 1987. Comparison of colony defense by European, hybrid (EXA) and mixed honey bee colonies. *Amer. Bee Jour.* 127: 842–845.

Collins, A. M. and T. E. Rinderer. 1991. Genetics of defensive behavior. In: *The "African" Honey Bee.* Ed. M. Spivak, D. J. C. Fletcher and M. D. Breed. Westview Press, Boulder, CO. Ch. 16. 309–328.

Crane, E. 1990. *Bees and Beekeeping.* Cornell Univ. Press. Ithaca, NY. 614 pp.

DeGrandi-Hoffman, G., A. M. Collins, J. H. Martin, J. O. Schmidt and H. G. Spangler. 1998. Nest defense behavior in colonies from crosses between Africanized and European honey bees. (*Apis mellifera* L.) (Hymenoptera: Apidae). *Jour. Insect Behav.* 11:37–45.

Dietz, A., R. Krell, and F. A. Eischen. 1985. Preliminary investigation on the distribution of Africanized honey bees in Argentina. *Apidologie.* 16:99–108.

Dietz, A. and C. Vergara. 1995. Africanized honey bees in temperate climates. *Bee World* 76(2):56–71.

De Jong, D. 1994. The African experience. *Bee Culture* 122(8):453, 456–461.

Fell, R. D. 1997. Insects: Hymenoptera (ants, wasps and bees). In: *Honey Bee Pests, Predators and Diseases*, 3rd Ed. Ed. R. A. Morse and K. Flottum. A. I. Root Co., Medina, OH. Ch. 9. 165–200.

Fletcher, D. J. C. 1991. Interdependence of genetics and ecology in a solution to the African bee problem. In: *The "African" Honey Bee.* Ed. M. Spivak, D. J. C. Fletcher and M. D. Breed. Westview Press, Boulder, CO. Ch. 4. 77–94.

Gentry, C., M. Wilburn, R. Rice and C. Sommer. 1985. *A manual for trainers of small scale beekeeping development workers.* Peace Corps, Washington, DC. 406 pp

Gonzalez, M. 1989. *Technicas modernas para la explotacion de la abeja Africanizada*. APES, Alajuela, Costa Rica. 70 pp.

Guzman-Novoa, E. and R. E. Page Jr. 1993. Backcrossing Africanized honey bee (*Apis mellifera* L.) queens to European drones reduces colony defensive behavior. *Ann. Entomol. Soc. Am.* 86:352–355.

Guzman-Novoa, E. and R. E. Page, Jr. 1994. Genetic dominance and worker interactions affect honey bee colony defense. *Behav. Ecol.* 5:91–97.

Guzman-Novoa, E. and R. E. Page, Jr. 1994. The impact of Africanized bees on Mexican beekeeping. *Amer. Bee Jour.* 134:101–106.

Guzman-Novoa, E. and R. E. Page, Jr., and M. K. Fondrk. 1994. Morphometric techniques do not detect intermediate and low levels of Africanization in honey bee (*Apis mellifera* L.) colonies. *Ann. Entomol. Soc. Am.* 87:507–515.

Hellmich, R. L. and T. E. Rinderer. 1991. Beekeeping in Venezuela. In: *The "African" Honey Bee*. Ed. M. Spivak, D. J. C. Fletcher and M. D. Breed. Westview Press, Boulder, CO. Ch. 20. 399–411.

Johnston, A. N. and J. O. Schmidt. 2001. The effect of Africanized honey bees (Hymenoptera: Apidae) on the pet population of Tuscon: a case study. *Amer. Entomol.* 47(2):98–103.

Krell, R., A. Dietz, and F. A. Eischen. 1985. A preliminary study on winter survival of Africanized and European honey bees in Cordoba, Argentina. *Apidologie* 16:109–118.

Murray, J. A. 1964. A case of multiple bee stings. *Jour. Med.* 10:249–251.

National Academy of Sciences (NAS). 1972. *Final report. Committee on the African Honey Bee*. PB-211 952. National Research Council, National Academy of Sciences. Tech. Info. Service, U.S. Dept. of Commerce, Springfield, VA. 95 pp.

Page R. E., and W. E. Kerr 1991. Honey Bee Genetics and Breeding. In: *The "African" Honey Bee*. Ed. M. Spivik, D. J. C. Fletcher and M. D. Breed. Westview Press, Boulder, CO: Ch. 8. 157–186.

Pesante, D., T. E. Rinderer, A. M. Collins, A. J. Broykin and S. M. Buco. 1992. Honey production in Venezuela: Effects of feeding sygar syrup on colony weight gains by Africanized and European colonies. *Apidologie* 23:545–552

Rinderer, T. E. 1985. Africanized honeybees in Venezuela: Honey production and foraging behavior. In: *Apiculture in Tropical Climates, Vol. 3*. IBRA, Cardiff, Wales. 112–116

Rinderer, T. E. 1988. Evolutionary aspects of the Africanization of honey bee populations in the Americas. In: *Africanized Honey Bees and Bee Mites*. Ed. G. R. Needham, R. E. Page, M. Delfinado-Baker and C. Bowman. Chichester, England. 13–29.

Rodriguez-Lainz, A., C. L. Fritz and W. R. McKenna. 1999. Animal and human health risks associated with Africanized honey bees. *J. Am. Vet. Med. Assoc.* 215:1799–1804

Schumacher, M. J., J. O. Schmidt and W. B. Egan. 1989. Lethality of "killer" bee stings. *Nature* 337:413.

Schumacher, M. J., J. O. Schmidt, N. B. Egen and J. E. Lowery. 1990. Quantity, analysis and lethality of European and Africanized honey bee venoms. *Amer. Jour. Trop. Med. Hyg.* 43:79–86.

Schmidt, J. O. 1992. Allergy to venomous insects. In: *The Hive and the Honey Bee*. Dadant and Sons, Inc, Hamilton, IL. Ch. 27. 1209–1269.

Schmidt, J. O. 1992. Doggie gone but not forgotton: lessons from a fatal bee attack. *Amer. Bee Jour.* 132:830.

Stort, A. C. 1974. Genetic study of the aggressiveness of two subspecies of *Apis mellifera* in Brazil. I. Some tests to measure aggressiveness. *Jour. Apic. Res.* 13:33–38.

Stort, A. C. 1975. Genetic study of the aggressiveness of two subspecies of *Apis mellifera* in Brazil. II. Time at which the first sting reached the leather ball. *Jour. Apic. Res.* 14:171–175.

Stort, A. C. 1975. Genetic study of the aggressiveness of two subspecies of *Apis mellifera* in Brazil. IV. Number of stings in the gloves of the observer. *Behav. Genet.* 5:269–274.

Stort, A. C. 1975. Genetic study of the aggressiveness of two subspecies of *Apis mellifera* in Brazil. V. Number of stings in the leather ball. *Jour. Kans. Entomol. Soc.* 48:381–387.

Stort, A. C. and L. E. Gonçalves. 1991. Genetics of defensive behavior II. In: *The "African" Honey Bee*. Ed. M. Spivak, D. J. C. Fletcher and M. D. Breed. Westview Press, Boulder, CO. Ch. 17. 329–356.

Smith, F. G. 1960. *Beekeeping in the Tropics*. Longmans, London.

Townsend, G. F. 1973. The African honey bee (*Apis mellifera adansonii*). *Apiacta* 8:105–106.

Villa, J. D. 1988. Defensive behavior of Africanized and European honey bees at two elevations in Colombia. *Jour. Apic. Res.* 27:141–145.

Wiese, H. 1985. *Nova Apiculture*. 6th ed. São Paulo, Brazil. 366 pp.

Wilkins, G. 1990. *Good Farmers: Traditional Agricultural Resource Management in Mexico and Central America*. Univ. Calif. Press, Berkeley, CA. 302 pp.

Chapter 12

Boreham, M. M. and D. W. Roubik. 1987. Population change and control of Africanized honey bees (Hymenoptera: Apidae) in the Panama Canal area. *Bull. Entomol. Soc. Amer.* 33:34–39.

Burgett, D. M. 1996. International aid and beekeeping: Is it working? *Bee World* 77:173–175.

Caron, D. M. 1987. Surviving with Africanized bees. *Amer. Bee Jour.* 127 (8):561–563.

Danka, R. G. Loper, J. D. Villa, J. L. Williams, E. A. Sugden, A. M. Collins and T. E. Rinderer. 1994. Abating feral Africanized honey bees to enhance mating control of European Queens. *Apidologie* 25:520–529.

Danka, R. G., T. E. Rinderer, A. M. Collins, and R. L. Hellmich. 1987. Response of Africanized honey bees (Hymenoptera: Apidae) to pollination-management stress. *Jour. Econ. Entomol.* 80:621–624.

Danka, R. G., J. D. Villa, and N. E. Gary. 1993. Comparative foraging distances of Africanized, European and hybrid honey bees (*Apis mellifera* L.) during pollination of cantaloupe (*Cucumis melo* L.). *BeeScience 3.* 16–21.

Danka, R. G., R. L. Hellmich and T. E. Rinderer. 1992. Nest usurpation, supersedure and colony failure contribute to Africanization of commercially managed European honey bees in Venezuela. *Jour. Apic. Res.* 31(3/4):119–123.

Danka, R. G., R. L. Hellmich, A. M. Collins, and T. E. Rinderer. 1990. Flight characteristics of foraging Africanzied and European honey bees (Hymenoptera: Apidae). *Ann. Ento. Soc. Amer.* 83(4):855–859.

DeGrandi-Hoffman, G. and S. S. Schneider. 2001. Worker behaviors in queenless Africanized honey bee colonies. *Proc. 2nd Inter. Conf. on African Honey Bees and Bee Mites.* Abst 15. To be published by A. I. Root Co., Medina, OH.

DeGrandi-Hoffman, G., J. C. Watkins, A. M. Collins, G. M. Loper, J. H. Martin, M. C. Arias and W. S. Sheppard. 1998. Queen developmental time as a factor in the Africanization of European honey bee (Hymenoptera: Apidae) populations. *Ann. Ento. Soc. Amer.* 91:52–58.

Dietz, A., J. F. Lietner, C. Vergara and M. Mejia. 1989. Effects of prolonged confinement in a refrigeration chamber on the survival of Africanized and European honey bees. *Amer. Bee Jour.* 129:815.

Duay, A. and D. De Jong. 1994. Variacion de la produccion de miel con abejas Africanizadas seleccionadas (*Apis mellifera scutellata*), no seleccionada and híbridas (*A. m. ligustica* x *A. m. scutellata*) en condiciones de campo en Brasil. *Proc. IV Cong Apiculture Iberolatinoamericano.* Rio Cuarto, Cordoba, Argentina. Com I. 7–10.

Duay, P. R. 1996. A technique for installing mating nuclei with Africanized honey bees. *Bee World* 77:212–216.

Fewell, J and J. Harrison. 2001. Differences in worker task phenotypes of African and European honey bees. *Proc. 2nd Inter. Conf on African Honey Bees and Bee Mites.* Abst 1. To be published by A. I. Root Co., Medina, OH.

Fletcher, D. J. C. 1978. The African bee *Apis mellifera adansonii* in Africa. *Ann. Rev. Entomol.* 23:151–171.

Gonçalves, L. S. 1974. The introduction of the African bees (*Apis mellifera adansonii*) into Brazil and some comments on their spread in South America. *Amer. Bee Jour.* 114:414–415, 419, 448–450 and 115:8–10, 24.

Gonçalves, L. S., A. C. Stout and D. De Jong. 1991. Beekeeping in Brazil. In: *The "African" Honey Bee.* Ed. M. Spivak, D. J. C. Fletcher and M. D. Breed. Westview Press, Boulder CO. Ch. 18. 334–346.

Guzman-Novoa, E., R. E. Page and D. Prieto-Merlos. 1998. Queen introduction, acceptance and survival in honey bee (Hymenoptera: Apidae) colonies of a tropical, Africanized region. *Jour. Econ. Entomol.* 91:1290–1293.

Hellmich, R. L., A. M. Collins, R. G. Danka, T. E. Rinderer. 1988. Influencing of mating of European honey bee queens in areas with Africanized honey bees (Hymenoptera: Apidae). *Jour. Econ. Entomol.* 81:796–799.

Hellmich, R. L. and G. D. Waller. 1990. Preparing for Africanized honey bees: evaluating control in mating apiaries. *Amer. Bee Jour.* 130:537–542.

Hellmich, R. L., J. Ibarra, M. Meija, T. E. Rinderer and N. A. Gutierrez. . 1993. Evaluating mating control of honey bee queens in an Africanized area of Guatemala. *Amer. Bee Jour.* 133(3):207.

Hellmich, R. L. and T. E. Rinderer. 1991. Beekeeping in Venezuela. In: *The "Africanized" Honey Bee.* Ed. M. Spivik, D. J. C. Fletcher and M. D. Breed. Westview Press, Boulder, CO. Ch. 20. 399–411.

Kerr, W. E. 1984 Historia parcial de ciencia apicola no Brasil. In: *Proc. V Cong. Bras. Apic. y III Cong. Iberolatinoamericano.* Ed. L. S. Conçalves, A. E. E. Soares, D. De Jong, J. Steiner, M. R. Martinho and N. Message. Imprensa Univ., Vicosa, Minas Gerais, Brazil. 47–60.

Kerr, W. E. and M. Cabeda. 1985. Introuçao de abelhas no territorio. Federal de Fernando de Noronha. *Ciencia e Cultura.* 37:467–471.

Kerr, W. E., L. S. Gonçalves, L. F. Blotta and H. B. Maciel. 1972. Biologia comparada entre as abelhas italianas (*Apis mellifera ligustica*), africana (*Apis mellifera adansonii*) e suas híbridas. *1º Bras. Cong. Apic.* Florianópolis, Santa Catarina, Brazil. 151–185.

Kigatiira, K. L. 1988. Amalgamation in tropical honey bees. In: *Africanized Honey Bees and Bee Mites.* Ed. G. R. Needham, R. E. Page, M. Delfinado-Baker, C. Bowman. Ellis Horwood Ltd. Chichester, England. 61–71.

Krell, R., A. Dietz, and F. A. Eischen. 1985. A preliminary study on winter survival of Africanized and European honey bees in Cordoba, Argentina. *Apidologie* 16:109–118.

Lee, P. C. & M. L. Winston 1987. Effects of reproductive timing and colony size on the survival, offspring colony size and drone production in the honey bee (*Apis mellifera*). *Ecol. Entomol.* 12:187–195.

Loper, G. M. and R. G. Danka. 1991. Pollination tests with Africanized honey bees in southern Mexico, 1986–88. *Amer. Bee Jour.* 131(3):191–193.

Loper, G. M. 1997 Overwinter losses of feral honey bee colonies in south Arizona 1992–1997. *Amer. Bee Jour.* 137 (6):446–448.

Loper, G. M. and M. M. Fierro. 1991. Use of drone trappings and drone releases to influence matings of European queens in an Africanized honey bee area (Hymenoptera:Apidae). *Jour. Apic. Res.* 30(3/4):119–124.

Loper, G. M. and E. A. Sugden 1992. Local Africanized honey bee abatement using drones as insecticide vectors in Costa Rica. *Amer. Bee Jour.* 132(12):847.

Portugal-Araújo, V de 1971. The central African bee in South America. *Bee World* 52:116–121.

National Academy of Sciences (NAS). 1972. *Final Report, Committee on the Africcan Honey Bee.* PB-211 952. National Research Council, National Academy of Sciences. Tech. Info. Service, U.S. Dept. of Commerce, Springfield, VA. 95 pp.

Nightingale, J. 1976. Traditional beekeeping among Kenya tribes and methods proposed for improvement and modernisation. In: *Apiculture in Tropical Climates*, Ed. E. Crane. IBRA, Cardiff, Wales. 15–22.

Oldroyd, B., A. Smokenski, S. Lawler, A. Estoup and R. Crozier. 1995. Colony aggregations in *Apis mellifera* L. *Apidologie* 26:119–130.

Otis, G. W., M. L. Winston, and O. R. Taylor 1981. Engorgement and dispersal of Africanized honeybee swarms. *Jour. Apic. Res.* 20:3–12.

Otis, G. W. 1982. Population biology of Africanized honey bee. In: *Social Insects in the Tropics. Proc. Sym. I.U.S.S.I.* Mexico. 209–219.

Ramirez, W. 1993. Swarming of the African bee in the central valley of Costa Rica, Central America. *Amer. Bee Jour.* 134(4):283–284.

Ratnieks, F., M. Piery and L. Quadriello. 1991. The natural nest of the Africanzied honey bee near Tapachula, Chiapas, Mexico. *Can. Entomol.* 123:353–359.

Rinderer, T. E., R. L. Hellmich, R. G. Danka and A. M. Collins. 1985. Male repoductive parasitism: a factor in the Africanization of European honey-bee populations. *Science* 228:1119–1121.

Rinderer, T. E. and A. M. Collins. 1991. Foraging behavior and honey producion. In: *The "African" Honey Bee.* Ed. M. Spivik, D. J. C. Fletcher and M. D. Breed. Westview Press, Boulder, CO. Ch. 12. 219–241.

Rinderer, T. E. and A. M. Collins. 1991. The Process of Africanization. In: *The "African" Honey Bee.* Ed. M. Spivik, D. J. C. Fletcher and M. D. Breed. Westview Press, Boulder, CO. Ch. 5. 95–118.

Roubik, D. W. and M. M. Boreman. 1990. Learning to live with Africanized honey bees. *Interciencia* Caracas, Venezuela. 15:146–153.

Roubik D. W. 1995. Pollination of cultivated plants in the tropics. FAO *Agric. Bull. 118.* 196 pp.

Roubik, D. W. 1988. An overview of Africanized honey bee populations: Reproduction, diet, and competition. In: *Africanized Honey Bees and Bee Mites.* Ed. G. Needham, M. Delfinado-Baker, R. Page, C. Bowman. Ellis Horwood Limited. Chichester, England. 45–54

Roubick, D. W. 1989. *Ecology and Natural History of Tropical Bees.* Cambridge Univ. Press, NY. 514 pp.

Roubick, D. W. 1995. *Pollination of cultivated plants in the tropics.* FAO Agric. Bull. 118. 196 pp.

Schiff, N. M., W. S. Sheppard, G. M. Loper and H. Shimanuki. 1994. Genetic diversity of feral honey bee (Hymenoptera: Apidae) populations in the southern United States. *Ann. Entomol. Soc. Amer.* 87(6):842–848.

Seeley, T. D. 1978. Life history strategy of the honey bee. *Oecologia* 32:109–118.

Seeley, T. D. 1985. *Honey Bee Ecology.* Princeton Univ. Press, Princeton, NJ. 201 pp.

Severson, D. W. 1984. Swarming behavior of the honey bee. *Amer. Bee Jour.* 124:204–210, 230–232.

Sheppard, W. S. 1989. A history of the introduction of honey bee races into the United States. *Amer. Bee Jour.* 129:664–667.

Smith, F. G. 1961. The races of honey bees in Africa. *Bee World* 42:225–260.

Spivak, M. 1991. The Africanization process in Costa Rica. In: *The "African" Honey Bee.* Ed. M. Spivak, D. J. C. Fletcher and M. D. Breed. Westview Press, Boulder, CO. Ch. 7. 137–155.

Spivak, M., E. Erickson. 1992. Do measurements of worker cell size reliably distinguish Africanized from European honey bees (*Apis mellifera* L.). *Amer. Bee Jour.* 132(4):252–255.

Taylor, O. R. 1977. Past and possible future spread of Africanized honey bees in the Americas. *Bee World* 58:19–30.

Taylor, O. R. 1985. African bees: Potential impact in the United States. *Bull. Entomol. Soc. Amer.* 31:15–24.

Taylor, O. R. 1988. Ecology and economic impact of African and Africanized honey bees. In: *Africanized Honey Bees and Bee Mites.* Ed. G. R. Needham, R. E. Page, M. Delfinado-Baker, and C. Bowman. Ellis Horwood Ltd., Chichester, England. Ch. 3. 29–40.

Taylor, O. R. 2000. The African honey bee in the Americas. *Proc. 36th Internat. Congr. Apimondia.* 202.

Vergara, C., A. Dietz, A. Perez. 1989. Usurpation of managed honey bee colonies by migratory swarms in Tabasco, Mexico. *Amer. Bee. Jour.* 129:824–825.

Visscher, P. K., T. D. Seeley. 1982. Foraging strategy of honey bee colonies in the temperate deciduous forest. *Ecology* 63:1970–1801.

Wiese, H. 1977. Apiculture with Africanized bees in Brazil. In *African Bees.* Ed. D. J. C. Fletcher. Apimondia, Pretoria, South Africa. 67–76. Also in *Amer. Bee Jour.* 117:166–170.

Winston, M. L. *Biology of the Honey Bee.* Harvard University Press, Cambridge, MA.

Winston, M. L. 1992. The Africanized honey bee. *Annual Rev. Entomol.* 37:173–193.

Winston, M. L. 1991. The Inside story: Internal Colony Dynamics of Africanized Bees. In: *The "African" Honey Bee.* Ed. M. Spivik, D. J. C. Fletcher and M. D. Breed. Westview Press, Boulder, CO. Ch. 7. 201–212.

Winston, M. L., G. W. Otis, and O. R. Taylor. 1979. Absconding behavior of the Africanized honey bee in South America. *Jour. Apic. Res.* 18:85–94.

Winston, M. L., J. A. Dropkin and O. R. Taylor. 1981. Demography and life history characteristics of two honey bee races (*Apis mellifera*). *Oecologia* 48:407–413.

Winston, M. L. 1995. Killer Bee Killers. *Bee Culture* 123:677–678.

Villa, J. D. 1987. Africanized and European colony conditions at different elevations in Colombia. *Amer. Bee Jour.* 127:53–57.

INDEX

INDEX

A

abscond(s) 102, 172, 176
absconding 1, 2, 31, 37, 47, 51, 88, 89, 99, 102, 103, 112, 115, 145, 169, 170, 171, 172, 173, 175, 176, 177, 181
Africa 2, 4, 11, 25, 26, 27, 28, 29, 30, 31, 34, 36, 40, 45, 58, 147, 166, 170, 171, 173, 175, 181
afterswarm(s) 171, 174
Agroyungas 117
Alabama 22
alarm odor (chemical) 74, 80, 120
alarm pheromone 138, 147, 148, 152, 160, 164
allergic response 150
allozymes 42, 107
Amazon River 13
Andes mountains 13, 14, 117, 118, 119, 122, 126, 140, 141, 169, 170
apiary site 156, 157, 158
Apisbol 117, 120, 127
Apis cerana 27
A. dorsata 27
A. florea 27
A. indica 27
Apis mellifera 11, 27, 178, 180
A. m. adansonii 4, 30, 32, 39, 54, 112, 134, 153, 165
A. m. brasiliensis 39
A. m. capensis 30, 32
A. m. carnica 34
A. m. caucasica 34
A. m. intermissa 34
A. m. ligustica 34
A. m. litorea 30
A. m. mellifera 34
A. m. monticola 30, 33, 153
A. m. scutellata 4, 30, 31, 32, 33, 34, 39, 44, 54, 152, 153, 165, 168, 177
A. m. unicolor 30
"American" bee 50
Apistan 23
appropriate technology 168
Argentina 13, 14, 25, 42, 43, 115, 168, 169, 176, 178
Arizona 13, 18, 20, 21, 26
army ants 158, 176
Australia 178

B

backcrossing 147
bait hives 15, 17, 20, 23, 56, 61, 64, 113
barrier 52, 56, 57, 58, 60, 61, 62, 64
Bee Kit 18
bee products 2, 120
Bee Regulated Zone (BRZ) 59, 60, 61, 62
bee space 1, 153, 167
"Bee Swat" team 58, 79, 85, 86, 98
beeswax 31
behavior 45, 46, 47, 100, 108,
Belize 49, 129–133, 152, 180
Bolivia 13, 57, 104, 109, 111, 129, 135, 139, 140, 141, 155, 158, 168, 169
Brazil 2, 4, 5, 6, 7, 9, 11, 12, 13, 14, 30, 32, 33, 35, 42, 43, 45, 58, 64, 66, 71, 72, 74, 111, 112, 115, 126, 127, 133–139, 140, 144, 146, 148, 149, 150, 151, 158, 161, 165, 166, 171, 172, 176, 178, 179, 181
Brazilian Apicultural Congress 6, 7
Brazilian bee 1, 6, 56, 165
brood 49
brood rearing 161, 170, 171, 178
Burgett, D. M. 14, 130, 131

C

California 13, 18, 21, 22, 23, 24, 26
California, Lost Hills 17
campesino 8, 67, 68, 80, 113, 117, 121, 122, 127, 135, 144, 168, 181
Canada 4, 25, 79, 113, 124, 133
Canadian barrier 62
Caribbean 4
Carniolan bee 34, 179
Caucasian bee 34, 135
cell cappings 48
cell size 46, 100, 102
Chile 4, 14, 113, 115, 118
China 27, 178
Chiriquí 66, 67, 70, 71, 73, 75, 85, 89, 90
cluster 170
coca (cocaine) 94, 117, 121, 122
colonization 2, 3, 14, 24, 44, 46, 49, 92, 137, 150, 154, 157, 164
colonizing 167, 176, 177, 178, 179, 182
Columbia 65, 85, 104, 144, 149, 169
Copan 8
Costa Rica 36, 50, 58, 64, 65, 66, 89, 93–110, 129, 146, 150, 169, 170, 171, 174, 175
coveralls 159
criolla bee 124, 125, 127

D

dance language communication 163, 164
dancing 5
DEET® 86
defensive 2, 4, 5, 12, 13, 30, 31, 32, 33, 34, 35, 45, 46, 47, 48, 49, 50, 79, 80, 87, 91, 94, 95, 96, 97, 98, 99, 100, 101, 104, 105, 106, 107, 112, 113, 122, 124, 125, 134, 136, 137, 138, 141, 143, 145, 147, 148, 149, 150, 151, 152, 153, 158, 159, 161, 163, 164, 169, 170, 177, 181
defensiveness 54, 56
De Jong, D. 7, 9, 77, 126, 134, 136, 137, 138, 139, 158, 161, 166
Denmark 124
desert bee 18
destruct-harvest 152–155, 156, 177
DNA 14, 28, 42, 43, 44, 45, 54, 124
drifting 35
Drone Congregation Areas 36
drone parasitism 35, 153
drones 100, 107, 171, 175, 178
drone semen 17
drone traps 60

E

ecotype 4, 44, 45, 176, 182
Ecuador 14, 111
European foulbrood 112, 115
European Union (EU) 117, 122

F

FABIS 20, 41
FAO (United Nations) 117, 126, 139
feral 1, 4, 13, 18, 19, 20, 23, 24, 26, 33, 40, 43, 44, 46, 53, 54, 63, 64, 82, 85, 86, 89, 92, 93, 94, 98, 100, 104, 107, 110, 113, 122, 126, 127, 142, 144, 152, 157, 158, 167, 168, 170, 175, 177, 178, 179, 180, 181
fire ants 158
Florida 14, 16, 17, 22, 23
foraging 2, 37, 48, 49, 60, 108, 161, 162, 163, 164
founding effect 42
French Guiana 36, 76, 171, 172, 174

G

gene(s) 44, 53, 178
genetic 17, 21, 28, 36, 151

genetic barrier 58, 62
genetic drift 42
genetic makeup 3
genetic variation 179
German bee(s) 9, 12, 34, 35, 134, 136, 153, 166, 179
Germany 25
gleaning 162, 163, 164
gloves 160
Gonçalves 7, 134, 136, 138, 139, 158, 165
gorilla (as bee enemy) 55
Great Britain 129, 131
GTZ 67, 93
guard bees 151
guards 48
Guatemala 57, 58, 66, 129
Guyana 13

H

Hive and the Honey Bee 2, 29, 167
hive stands 85, 88, 101, 156, 158, 159, 164
hobby beekeeping 22, 24
hobbyist 63, 68, 70, 126, 144, 165
Honduras 8, 40
honey 165, 166, 171, 172, 175, 176, 178
honey export 73
honey house 72, 116
honey hunters 31, 32, 168, 177, 180
honey price 119, 144, 145
honey production 30, 31, 32, 33, 45, 54, 73, 75, 83, 86, 87, 89, 91, 99, 102, 105, 106, 112, 113, 114, 115, 117, 119, 124, 126, 130, 133, 135, 136, 137, 140, 141, 142, 143, 145, 151, 153, 154, 156, 161–163, 164, 165, 166, 167, 181
hybrid 26, 39, 42, 43, 44, 45, 64, 94, 104, 107, 108, 110, 135, 136, 139, 147, 149, 150, 176, 182
hybridization 14

I

Iberian bee 34
Illinois 22
Indian honey bee 27
insecticide curtain 57
interbreeding 52, 53
Italian bees 9, 11, 12, 25, 33, 34, 35, 78, 93, 106, 113, 115, 122, 124, 125, 126, 131, 134, 135, 138, 141, 144, 149, 153, 166, 167, 178, 179, 181

K

Kenya top bar hive (KTBH) 153, 154
Kerr, W. E. 11, 12, 14, 30, 33, 39, 42, 58, 112, 136, 166, 171,
killer bee(s) 1, 2, 4, 5, 7, 8, 18, 25, 39, 51, 81, 134
"killer" honey 54

L

LD_{50} 150
leather glove test 148
length of life 49
Lost Hilla, CA 17
Louisiana 17, 149

M

Maine 22
manipulation cloths 160, 161
marijuana 131
mating 35, 36, 53, 54, 153, 175
Maya 8, 129, 131, 132, 133
megaswarm 89, 175, 177
Mexico 11, 15, 16, 23, 35, 36, 43, 49, 50, 52, 53, 59, 60, 61, 62, 64, 78, 98, 129, 131, 139, 171, 176, 178
MFPMEFE, Trinidad & Tobago 141, 142, 143, 144
Miel Mañana 69, 70, 76
migration 139

migratory 31, 37, 114, 118, 119, 120, 175
Ministry of Agriculture (MAG), Costa Rica 94, 101, 105, 110
Ministry of Agriculture (MIDA), Panama 66, 74, 82, 83, 85, 89
Mississippi 22
mites 34, 62, 179
morphology 28, 30, 39
morphometric(s) 29, 39, 40, 41, 44, 46
morphometric measurement(s) 14, 17, 18, 20, 21, 24, 87, 125
Morse, R. A. 60, 137
movies about bees 51
moving colonies 108, 120
mtDNA 14, 34, 42, 43, 44, 45, 54, 107
mtDNA analysis 21
myths 51–57, 64, 179, 183

N

National Beekeeping Congress, Bolivia 126
National Commission, Panama 83, 85, 101
National Honey Festival 114, 115, 116
NAS Committee report 1, 5, 6, 13, 35, 51, 56, 58, 63, 134, 137, 138, 148
Nasonov pheromone 20
nectar 165, 166, 170, 171, 172, 178
nectar flow 162
neotropical 39, 50
nest cavities 37
Nevada 13
New Mexico 13, 20, 21
Nogueira-Neto, P. 12, 32, 57, 134
Nosema disease 34
"numbers" game 36, 54

O

overwinter 25

P

package bees 62
Page, R. 21, 23, 43, 53, 147, 166
Panama 15, 33, 36, 40, 44, 49, 50, 52, 57, 58, 59, 62, 63, 64, 65–75, 78, 79–92, 93, 98, 101, 129, 144, 145, 146, 150, 158, 162, 171, 175
Panama Canal 58, 62, 63, 66, 79, 85, 86, 90, 150, 151, 167
parade 48, 74
Paraguay 13, 111, 168
Partners of the Americas 67
Peace Corp 105, 120, 132, 133, 154, 157
personal protection 159, 164
Peru 13, 14, 25, 64, 111, 139–141, 176
pheromone 60
pheromone-baited traps 142
Philippines 27
piñata traps 20, 130
pioneer swarms 182
poison bait 56
pollen 2, 33, 37, 90, 91, 105, 119, 127, 143, 144, 151, 163, 170, 171, 172, 175, 176
pollination 2, 21, 24, 96, 97, 99, 107, 108, 109, 110, 114, 115, 145, 164
pollinators 181
propolis 2, 113, 119, 124
protein(s) (allozymes) 14, 42
Puerto Rico 16, 17
punic bee 29, 30

Q

quarantine 18, 21, 24
queen breeder 46
queen breeding 134, 137, 142
queen cells 172
queen excluder 12, 60
queen parasitism 35
queen pheromone 36

queen rearing 12, 17, 24, 70. 72, 89, 90, 93, 94, 95, 100, 105, 106, 113, 131, 172
queens 11, 35, 36, 47, 113, 115, 124, 126, 131, 137, 138, 142, 144, 150, 153

R

race (of bee) 2, 4, 27, 28, 29, 30, 32, 35, 39, 50, 107, 147, 161
Reader's Digest 7
requeen(ing) 47, 53, 60, 63, 89, 91, 92, 94, 95, 97, 102, 104, 106, 113, 131, 141
Rinderer, T. E. 17, 26, 44, 139, 145, 146, 147, 150, 154, 162, 163, 166, 171, 178
robbing 116, 152
Roubik, D. 33, 37, 44, 87, 90, 109, 163, 175
royal jelly 2, 90, 119
running behavior 48, 49, 163
rustic hive(s) 60, 61, 68, 102, 118, 119, 123, 137, 154

S

Sahara 34
SARH 59, 61
screw worm fly 59, 60, 95
smoke 48
smoker (Brazilian) 87, 113, 160
South Africa 31
Spain 34
Spanish bees 11
species 27, 28, 39
Spivak, M. 39, 46, 98, 100–104, 108, 140, 169, 174
sting(ing) 1, 2, 3, 7, 8, 18, 19, 20, 21, 22, 31, 48, 54, 58, 74, 75, 76, 77, 79, 80, 81, 82, 83, 84, 85, 88, 91, 97, 98, 99, 101, 103, 120, 132, 134, 135, 137, 141, 144, 146, 147, 148, 149, 150, 151, 156, 158, 159, 160, 161, 163, 164, 165
stingless bees 11, 110, 113, 116, 134, 167, 180
subspecies 39
subtropical 178, 179, 181
supersedure 47
Surinam 13, 76, 77
swarm amalgamation 37
swarm(ing) 1, 2, 8, 12, 17, 19, 20, 21, 31, 34, 36, 44, 47, 56, 58, 60, 61, 63, 66, 68, 75, 79, 85, 89, 91, 92, 99, 101, 102, 103, 112, 113, 115, 118, 122, 123, 127, 134, 137, 141, 142, 145, 163, 167, 168, 169, 170, 171, 172, 173, 174, 175, 176, 178, 180, 181
swarm traps 20
"swat" team 142

T

Tabora 12
Tanzania 11, 12, 31
Taylor, O. R. 12, 24, 36, 39, 40, 44, 46, 51, 64, 75, 83, 98, 99, 138, 144, 169, 170, 178
Tehuantepec 59, 61
Texas 13, 14, 16, 18, 19, 20, 24, 26, 52, 64, 178
The Hive and the Honey Bee 2, 29, 167
thelytoky 32
thorax 30
TIME 7
toxic reaction 150
trachael mites 152
transition zone 14, 25, 26, 33, 43, 178
trap line 20, 22, 61
traps 60
Trinidad & Tobago 16, 141, 144, 146

U

Uruguay 14, 42, 126
USAID 117, 121, 129

USDA 2, 6, 10, 14, 15, 16, 17, 20, 22, 23, 26, 35, 41, 44, 46, 51, 59, 61, 81, 83, 95, 107, 108, 131, 135, 139, 145, 147, 148, 150, 162, 163, 166, 171, 178
usurpation 35, 53, 177

V

Varroa 20, 21, 23, 41, 53, 61, 89, 94, 115, 181
Venezuela 13, 18, 35, 43, 46, 53, 83, 85, 141, 144, 145, 146, 147, 149, 150, 163, 166, 171, 176
venom 55
Virginia 22
Virgin Islands 16
VOCA 14
von Frisch 5

W

website 21, 26, 55
western (European) honey bee 27
West Virginia 22
wing beat frequency 46
Winston, M. L. 2, 25, 36, 44, 57, 62, 76, 77, 139, 162, 166, 169, 172, 179
worker size 46
world record honey harvest 31

Y

Yucatan 35, 43, 44, 64, 131